Continuous Optimization for Data Science

Continuous Optimization for Data Science

Moshe Haviv

The Chinese University of Hong Kong, Shenzhen, China
The Hebrew University of Jerusalem, Israel

World Scientific

NEW JERSEY · LONDON · SINGAPORE · BEIJING · SHANGHAI · TAIPEI · CHENNAI

Published by

World Scientific Publishing Co. Pte. Ltd.

5 Toh Tuck Link, Singapore 596224

USA office: 27 Warren Street, Suite 401-402, Hackensack, NJ 07601

UK office: 57 Shelton Street, Covent Garden, London WC2H 9HE

Library of Congress Control Number: 2025020922

British Library Cataloguing-in-Publication Data
A catalogue record for this book is available from the British Library.

CONTINUOUS OPTIMIZATION FOR DATA SCIENCE

ISBN 978-981-12-9919-3 (hardcover)
ISBN 978-981-98-0150-3 (paperback)
ISBN 978-981-12-9920-9 (ebook for institutions)
ISBN 978-981-98-0087-2 (ebook for individuals)

For any available supplementary material, please visit
https://www.worldscientific.com/worldscibooks/10.1142/14041#t=suppl

Typeset by Stallion Press
Email: enquiries@stallionpress.com

Continuous Optimization for Data Science

Moshe Haviv

The Chinese University of Hong Kong, Shenzhen, China

The Hebrew University of Jerusalem, Israel

World Scientific

NEW JERSEY · LONDON · SINGAPORE · BEIJING · SHANGHAI · TAIPEI · CHENNAI

Published by

World Scientific Publishing Co. Pte. Ltd.

5 Toh Tuck Link, Singapore 596224

USA office: 27 Warren Street, Suite 401-402, Hackensack, NJ 07601

UK office: 57 Shelton Street, Covent Garden, London WC2H 9HE

Library of Congress Control Number: 2025020922

British Library Cataloguing-in-Publication Data
A catalogue record for this book is available from the British Library.

CONTINUOUS OPTIMIZATION FOR DATA SCIENCE

ISBN 978-981-12-9919-3 (hardcover)
ISBN 978-981-98-0150-3 (paperback)
ISBN 978-981-12-9920-9 (ebook for institutions)
ISBN 978-981-98-0087-2 (ebook for individuals)

For any available supplementary material, please visit
https://www.worldscientific.com/worldscibooks/10.1142/14041#t=suppl

Typeset by Stallion Press
Email: enquiries@stallionpress.com

Contents

Preface

Continuous optimization, sometimes referred to as nonlinear programming (NLP), is an old topic that deals with optimizing continuous functions. It has two main branches: unconstrained and constrained optimization. This text presents both theory and algorithms for solving such problems. Both branches are now more popular than ever and are being used in statistics, data science, and machine learning. Indeed, many of the examples given in the text are from the theory of statistics.

The text is suitable for third-year bachelor and first-year master students in statistics and data science, computer science, operations research, engineering, and mathematics. I believe that anyone who is interested in optimization should acquire the material given in the text. For those who wish to specialize in NLP, it can be used mainly as an introductory text, prior to moving to more advanced and more detailed texts such as [5].

A course in multi-variable calculus is a prerequisite for this text. The same can be said for linear algebra. A possible linear algebra text is [14], where Chapters 1–10 are required. On the other hand, no prior knowledge of linear programming is required. Sections 1.3.2, 1.3.3, 2.3.1, 2.3.2, 2.9, and Example 7 in Section 4.2.1 require previous exposure to parametric statistics, but they can be skipped without loss of continuity.

The text is comprised of three parts and in the following order: unconstrained optimization, constrained optimization, and linear programming (LP). Part I on unconstrained optimization is covered in Chapters 1 and 2. Chapter 1 examines single-variable functions, where conditions for optimality are set and algorithms for finding the optimal points are designed. In Chapter 2 we examine multi-variable functions, tackling similar problems. In particular, we introduce the steepest descent, Newton, and quasi-Newton algorithms. This chapter ends with the Expectation-Optimization

(EM) algorithm for parameter estimations in probabilistic models, and it can be skipped without loss of continuity.

Part II on constrained optimization is covered in Chapters 3 and 4. Chapter 3 deals with equality constraints, where the objective function is separable in the decision variables. Some illuminating examples dealing with queueing models are given. We end this chapter by deriving solutions to cases where all constraints are linear. The assumption of linear equality constraints simplifies the analysis considerably. Moreover, it paves the way for formulating and understanding a constrained optimization problem, now that the constraints are not necessarily linear. The approach taken here is quite different from those taken in most textbooks that usually do not deal separately with linear constraints. Chapter 4 discusses the case where no further assumptions are made on the objective and the constraint functions, but still under the assumption of equality constraints. In particular, conditions for optimality are stated. Sensitivity analysis is done, and a duality theorem of nonlinear programming is stated and proved. In Chapter 5, we add inequality constraints. Again, the optimality conditions are set, where the analysis stems from the case of equality constraints and uses the sensitivity analysis done for this case. Finally, a number of algorithms for solving constrained optimization problems are provided.

Part III on linear programming is covered in Chapters 6, 7, and 8. This part is self-contained and can be read independently of the rest of the text. It is put as Part III as I believe that, educationally, it is better to first study NLP and then switch to the special features of LP. Chapter 6 formulates the LP problem, while Chapter 7 gives basic solutions, states the simplex algorithm, and proves its finite convergence to the required optimal solution. Chapter 8, the concluding chapter, discusses sensitivity analysis. Other algorithms, such as those based on interior points, are beyond the scope of this book.

I was not alone in this project. Many thanks are due to those who assisted me. Yonatan Woodbridge was in charge of the solution sets of the first four chapters. Niv Brosh went through the text and spotted many points that called for correction. He also took care of the solution set for the exercises of Chapter 5. Zhenduo Wen and Ruohang Wei created the illuminating graphics. Mor Ben Ari, Anat Cohen and Yoseph Joffe were very helpful at the final stages of editing this book. Finally, Doron Haviv gave me a few ideas on what to include, as well as what not to include, in the book. Financial support by ISF grant 1512/19 is acknowledged.

Part I
Unconstrained Optimization

Chapter 1

Single-Variable Optimization

1.1 Introduction

Let $f(x) : R \to R$ be some function. We are interested in solving the problem $\min_{x \in R} f(x)$ and, in particular, in finding the argument where this minimization is attained (if it exists), denoted by $\arg\min_{x \in R} f(x)$. In principle, $\arg\min_{x \in R} f(x)$ is a set, and hence can be either empty, a single point, a finite number of points, or an infinite number of points. Since maximizing $f(x)$ is equivalent to minimizing $-f(x)$, for convenience, we deal mainly with the minimization problem.

Unless some mild conditions exist, this is a very hard problem. Specifically, even if one has a good guess as to the optimal point, how can one be sure that there is not a better point somewhere along the real-number axis? Thus, in many cases, an optimal solution, called a *global* solution, is too much to ask for, and one must resort to a search for *local* solution:

Definition

Definition 1.1. A point x^* is said to be a local minimal point for $f(x)$ if there exists some $\epsilon > 0$ such that $f(x^*) \leq f(x)$ for any x satisfying $|x - x^*| \leq \epsilon$.

Of course, a global minimum is also a local minimum, but not the other way around. The snag in this definition is of course that we say nothing about ϵ, which in principle can be (very) small. Moreover, it is possible that there are many local minima but no global minimum exists. In Figure 1.1, for example, a global maximum does not exist. Yet, in practice, things are

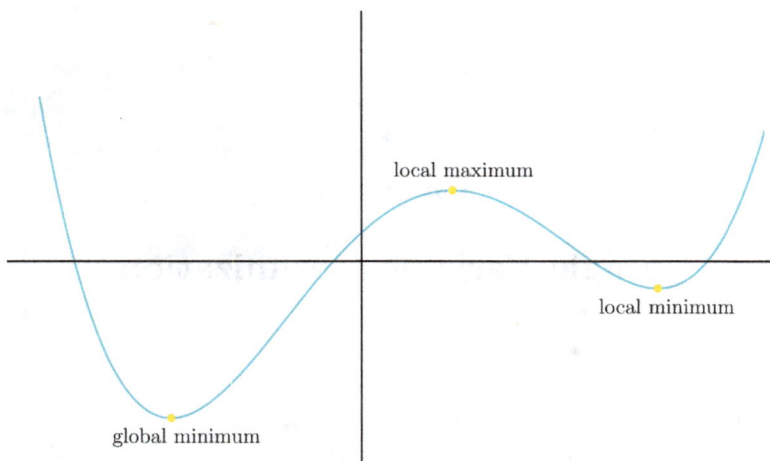

Figure 1.1. Local minimum and local maximum.

usually better. Normally, there are only a few local minima, each of which is a candidate to be the global minimum. Of course, the focus will then be on the point that receives the lowest value when the function $f(x)$ is applied to it.

The rest of this chapter deals with characterizing local minima and developing algorithms for finding them, or at least for finding good approximations of them. The importance of this search is twofold. Firstly, there are many single-variable optimization functions that emerge in practice in real life. Secondly, such functions are used as subroutines when multi-variable functions are optimized. Details on this approach are given in the next chapter. Note that the functions we deal with are differentiable: The fact that we refer to their derivatives by default assumes their differentiability.

1.2   First- and Second-Order Conditions

Using a Taylor series expansion for a function $f(x)$ around an arbitrary point x_0, we know that for any x,

$$f(x) = f(x_0) + f'(x_0)(x - x_0) + \frac{1}{2}(x - x_0)^2 f''(y), \qquad (1.1)$$

for some y that is a function of x and lies between x and x_0. From this, we infer that a necessary condition for x_0 to be a local minimum is that $f'(x_0) = 0$. Indeed, assume without loss of generality that $f(x_0) = 0$.

Then, if $f'(x_0) > 0$, it is possible to find $x < x_0$ that is close enough to x_0, where the right-hand side of the equality is negative. Note that this is the case regardless of the sign of $f''(y)$, since near x_0, $x - x_0$ is, relatively, much larger than $(x - x_0)^2$. A similar argument holds for $x > x_0$, where $f'(x_0) < 0$. Note also that the same necessary condition holds for the case where maximization is looked for. The condition $f'(x^*) = 0$ is known as the *first-order condition* (FOC) for optimization. At times, such a point is called *stationary* due to the fact that if we look at the function at x^* and ignore higher powers in the Taylor series expansion, it does not change in any direction as one moves away from x^*.

Clearly, this condition is not a sufficient condition for local optimization. In particular, it does not distinguish between minimum and maximum points. Inspect (1.1) again and assume that x_0 meets the FOC. Now the question of whether $f(x) > f(x_0)$ or not boils down to inspecting the sign of $f''(y)$. This sign coincides with the sign of $f''(x_0)$ if x is close enough to x_0 (unless $f''(x_0) = 0$, where the sign of $f''(y)$ is unknown). Thus, we conclude that the requirement $f''(x_0) \geq 0$ is a necessary condition for local minimization, while $f''(x_0) > 0$ is a sufficient one (once the FOCs are met, of course). This condition is called the *second-order condition* (SOC) for minimization. Note that the last two inequalities are reversed for maximization. To reiterate: SOCs are considered only at points that satisfy the FOCs.

Example 1 (Affine functions). Let $f(x) = ax + b$ for some constants $a \neq 0$ and b. As $f'(x) = a$, neither minimum nor maximum points exist.

Example 2 (Quadratic functions, see Figure 1.2). Let $f(x) = \frac{1}{2}ax^2 - bx + c$ for some constants $a \neq 0$ b and c. Clearly, $f'(x) = ax - b$. Then, a single point, $x = b/a$, satisfies the FOCs. Finally, $f''(x) = a$ for any x, in particular, $f''(b/a) = a$. Hence, b/a is a local minimum (maximum, respectively) point if $a > 0$ ($a < 0$, respectively).

In a special case of the above, let y_i, $1 \leq i \leq n$, be a sequence of real numbers.
Question: What is the value of x^*, where

$$x^* = \arg \min_x \sum_{i=1}^{n} (y_i - x)^2?$$

Let $f(x) = \Sigma_{i=1}^{n}(y_i - x)^2$. Then,

$$f(x) = \sum_{i=1}^{n} y_i^2 - 2x \sum_{i=1}^{n} y_i + nx^2.$$

$$f(x) = \frac{1}{2}ax^2 - bx + c$$

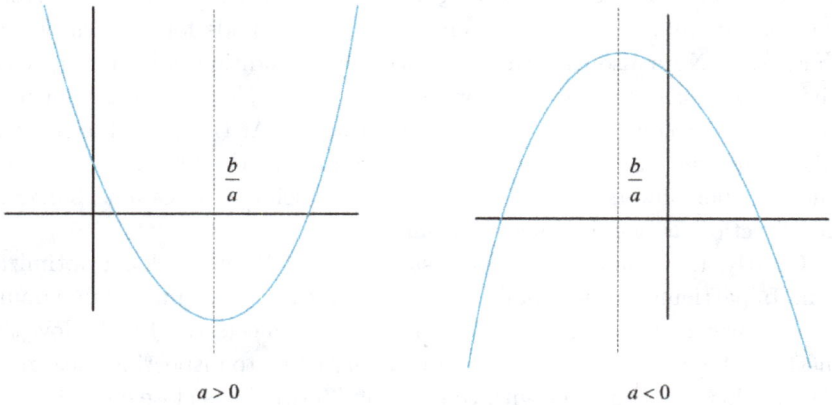

Figure 1.2. Local minimum and local maximum of quadratic functions.

Hence, $f'(x) = -2\Sigma_{i=1}^n y_i + 2nx$, and it equals zero at

$$x = \frac{\Sigma_{i=1}^n y_i}{n}, \tag{1.2}$$

namely, at the arithmetic mean of the sequence y_i, $1 \le i \le n$, usually denoted by $\bar{y}$. Finally, $f''(x) = 2n > 0$. Hence, this is a minimum point. For some other optimization problems of this type, which lead to other such solutions, see Exercise 1.1 and Section 6.2.1 below.

Exercise 1.1

Let $Y_1, Y_2, \ldots, Y_n$ be a series of n numbers. Let $\frac{1}{n}\Sigma_{i=1}^n Y_i$, $(\Pi_{i=1}^n Y_i)^{1/n}$, and $n/\Sigma_{i=1}^n \frac{1}{Y_i}$ be their arithmetic, geometric, and harmonic means, respectively. Note that for the existence of the geometric mean, one needs to assume further that $Y_i \ge 0$, $1 \le i \le n$.

1. Prove that the arithmetic mean of the series minimizes, with respect to y, the function

$$\sum_{i=1}^n (Y_i - y)^2.$$

2. Assume further that $Y_i > 0$, $1 \leq i \leq n$. Prove that the geometric mean of the series minimizes, with respect to y, the function

$$\sum_{i=1}^{n} (\log Y_i - \log y)^2.$$

3. Prove that the harmonic mean of the series minimizes, with respect to y, the function

$$\sum_{i=1}^{n} \left(\frac{1}{Y_i} - \frac{1}{y} \right)^2.$$

Example 3 (Cubic functions, see Figure 1.3). Let $f(x) = \frac{1}{3}ax^3 + \frac{1}{2}bx^2 + cx + d$ for some constants $a \neq 0, b, c$, and d. Then, $f'(x) = ax^2 + bx + c$ and $f''(x) = 2ax + b$. There are three possibilities here. The first is when $b^2 < 4ac$ and then no point satisfies the FOC for optimality, and the second is when $b^2 = 4ac$ and then only one point, $-b/2a$, does. In this case, $f''(-b/2a) = 0$, and therefore it is not clear whether this is a minimum, a maximum, or neither. Yet, some algebra shows that in this case

$$f'(x) = a \left(x + \frac{b}{2a} \right)^2,$$

and hence the function is monotone, either non-increasing or non-decreasing, depending on the sign of a. Thus, this is neither a minimum nor a maximum point, and in particular, we infer that the FOCs are not sufficient for proving the existence of either of these extreme points. Points that satisfy the FOCs but are not extreme points are referred to as *reflection points*. The third and final possibility is when $b^2 > 4ac$ and then two points satisfy the FOCs $f'(x) = 0$. They are, of course,

$$x_{1,2} = \frac{-b \pm \sqrt{b^2 - 4ac}}{2a}.$$

It is easy to see that

$$f''(x_{1,2}) = \pm\sqrt{b^2 - 4ac},$$

where one of them is a local minimum while the other is a local maximum. If $a > 0$, the minimum is at the positive root, while if $a < 0$, the minimum is at the negative root.

$$\frac{1}{3}ax^3 + \frac{1}{2}bx^2 + cx + d$$

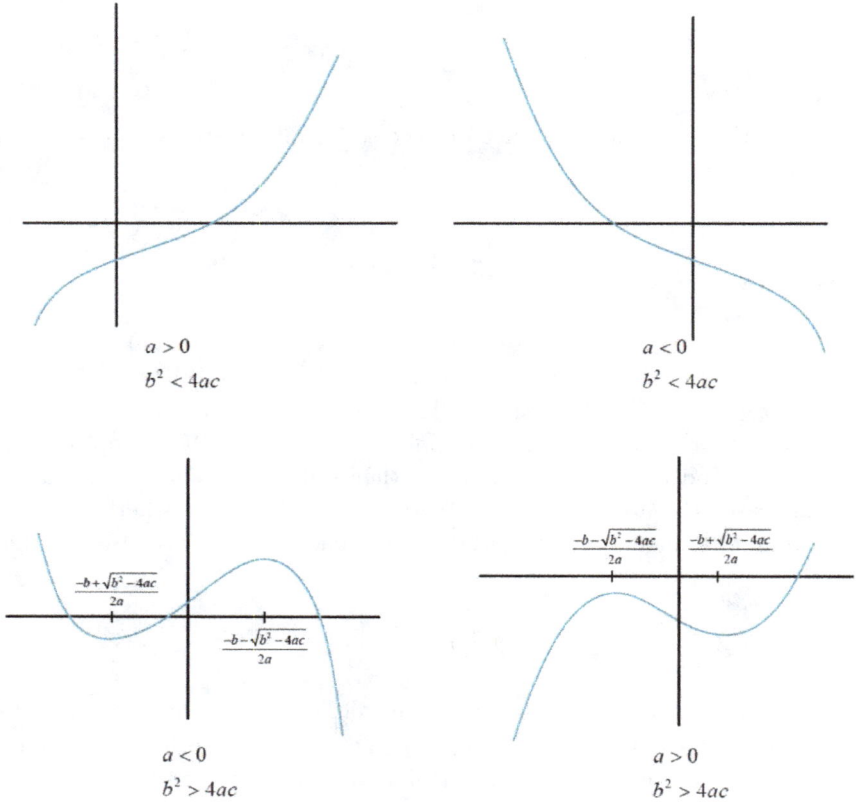

Figure 1.3. Local minima and local maxima of cubic functions.

1.3 Convex and Concave Functions

Inspecting the SOCs, one can observe that if $f''(x) > 0$ for all $x \in R$, then this condition automatically holds for any point we look at. Put differently, if this condition is met, all one needs to do is to consider the FOCs to find out whether or not minimization is attained. Moreover, and no less importantly, from (1.1) one can infer that the minimization is global and not only local.

> **Definition**
>
> **Definition 1.2.** For an interval X, a real function $f(x) : X \to R$ is said to be strictly convex (respectively, convex) if for all $x \in X$, $f''(x) > 0$ (respectively, $f''(x) \geq 0$). It is called strictly concave (respectively, concave) if $-f(x)$ is strictly convex (respectively, convex).

In convex functions one usually look for maxima, while for concave function the issue is maximization. It is possible to see that if $f(x)$ is convex, with $f(x) > 0$ and $f'(x) < 0$ for all $x \in X$, then $1/f(x)$ is convex too.

We next state a few conditions which individually are met if and only if a function is convex.

> **Theorem**
>
> **Theorem 1.1.** *The following properties are equivalent:*
>
> 1. $f''(x) \geq 0$ *for any x.*
> 2. *For any pair of x and y, $f(y) \geq f(x) + f'(x)(y - x)$ (see Figure 1.4(a)).*
> 3. *For any pair of x and y, and for any $0 \leq \alpha \leq 1$,*
>
> $$f(\alpha x + (1 - \alpha)y) \leq \alpha f(x) + (1 - \alpha)f(y)$$
>
> *(see Figure 1.4(b)).*
> 4. *For any x and any non-negative pair y and z, $f(x + z) - f(x) \leq f(x + y + z) - f(x + y)$ (see Figure 1.5).*

Proof. The equivalence between the first three properties is a special case of functions $f(x) : R^n \to R$ studied later in Theorem 2.1; see the full proof there. In any case, we would like to note that the equivalence between properties 1 and 2 is immediate from (1.1). The fourth property is more specific for functions $f(x) : R \to R$, and we turn to it first. Next, we prove the equivalent between properties 1 and 4. Suppose property 4 holds. Dividing both sides of the inequality

$$f(x + z) - f(x) \leq f(x + y + z) - f(x + y)$$

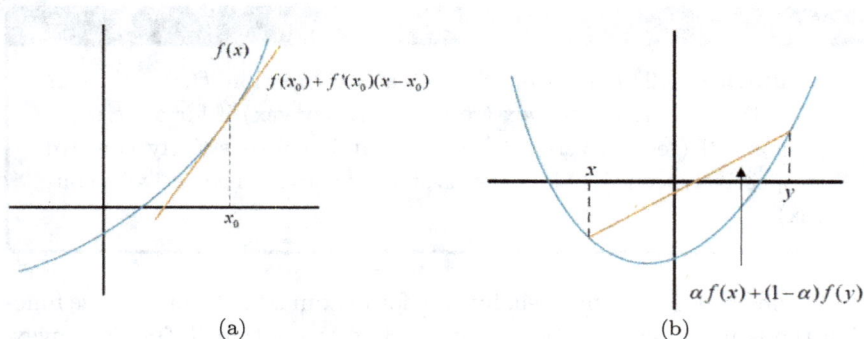

(a) (b)

Figure 1.4. Convex functions.

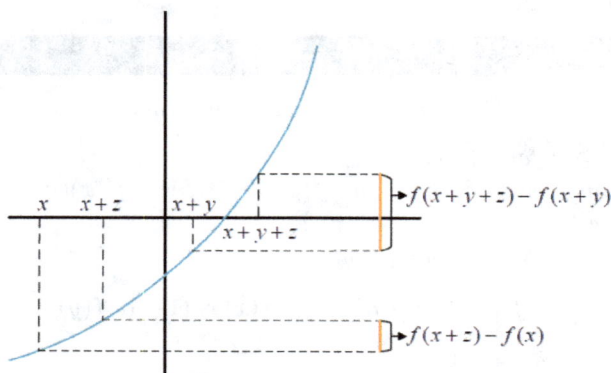

Figure 1.5. A convex function.

by z, which is positive, and taking the limit when z goes to zero, implies that $f'(x) \le f'(x+y)$. This leads to property 1. For the converse, assume that property 1 holds. Observe that

$$f(x+z) - f(x) = \int_{t=0}^{z} f'(x+t)\,dt \quad \text{and}$$

$$f(x+y+z) - f(x+y) = \int_{t=0}^{z} f'(x+y+t)\,dt.$$

Since property 1 says that the function $f'(x)$ is non-decreasing, we get that $f'(x+t) \le f'(x+y+t)$, $0 \le t \le z$. This completes the proof. $\square$

We claim without proving it that in the case where $f(x)$ is strictly convex, the above inequalities hold as long as $x \ne y$, $z = 0$, and $0 < \lambda < 1$.

Remark. The third property given above can be generalized as follows. Let $x_1, x_2, \ldots, x_n$ be a series of real numbers and let $\alpha_1, \alpha_2, \ldots, \alpha_n$ be non-negative numbers that sum up to 1. If $f(x)$ is convex, then

$$f\left(\sum_{i=1}^{n} \alpha_i x_i\right) \le \sum_{i=1}^{n} \alpha_i f(x_i). \tag{1.3}$$

Proof. This can be proved for by induction. Indeed, it holds for the case of n (by definition). Then, using the already known case when two point are involves, where the two points are $\sum_{i=1}^{n-1} \frac{\alpha_i}{1-\alpha_n} x_i$ and x_n, we have

$$f\left(\sum_{i=1}^{n} \alpha_i x_i\right) = f\left((1-\alpha_n)\sum_{i=1}^{n-1} \frac{\alpha_i}{1-\alpha_n} x_i + \alpha_n x_n\right)$$

$$\le (1-\alpha_n)f\left(\sum_{i=1}^{n-1} \frac{\alpha_i}{1-\alpha_n} x_i\right) + \alpha_n f(x_n).$$

As $\Sigma_{i=1}^{n-1} \frac{\alpha_i}{1-\alpha_n} = 1$, we invoke the induction hypothesis, and hence conclude that the right-hand side of the inequality above is less than or equal to

$$(1-\alpha_n)\sum_{i=1}^{n-1} \frac{\alpha_i}{1-\alpha_n} f(x_i) + \alpha_n f(x_n) = \sum_{i=1}^{n-1} \alpha_i f(x_i) + \alpha_n f(x_n) = \sum_{i=1}^{n} \alpha_i f(x_i),$$

as required. $\qquad\square$

The inequality stated in (1.3) is known in probability theory as the *Jensen inequality*, and it states the following: Let X be a random variable. Then, for a convex function $f : R \to R$, $f(\mathrm{E}(X)) \le \mathrm{E}(f(X))$, where the operation $\mathrm{E}(\cdot)$ returns the expected value. In this case, α_i stands for the probability that the random variable X takes the value of x_i, $1 \le i \le n$.

Remark. The fourth property stated above for convex functions is usually referred to as *increasing marginal costs* or *diseconomies of scale*. Indeed, for such cost functions, the larger the original quantity one possesses, the greater the effect that adding a constant to the input has on the total cost.

Example. For some positive constant a, consider the cost function $f(x) = 1/(a - x)$, $0 \le x < a$. It is easy to see that $f'(x) = 1/(a - x)^2$ and that $f''(x) = 2/(a - x)^3$, $0 \le x < a$. Thus, we have a monotone increasing convex function. One can observe that as x approaches the bound of a, the total cost increases. More than that: The rate of this increase, increases.

We next list a number of properties of convex functions. These are easily proven and hence no further details are given.

> **Theorem**
>
> **Theorem 1.2.** *Suppose that $f(x)$ and $g(x)$ are two convex functions. Then,*
>
> 1. *$af(x)$ is a convex function for any $a \geq 0$ and concave for any $a \leq 0$.*
> 2. *For any constant C, the set of points $\{x | f(x) \leq c\}$ is a closed interval (see Figure 1.6).*
> 3. *$f(x) + g(x)$ is convex function.*

1.3.1 *The inequality of the means[a]*

> **Definition**
>
> **Definition 1.3.** For n real numbers x_i, $1 \leq i \leq n$, their arithmetic mean is defined as $\frac{1}{n}\sum_{i=1}^{n} x_i$. In the case where additionally $x_i \geq 0$, $1 \leq i \leq n$, their geometric mean is defined as $(\Pi_{i=1}^{n} x_i)^{\frac{1}{n}}$.

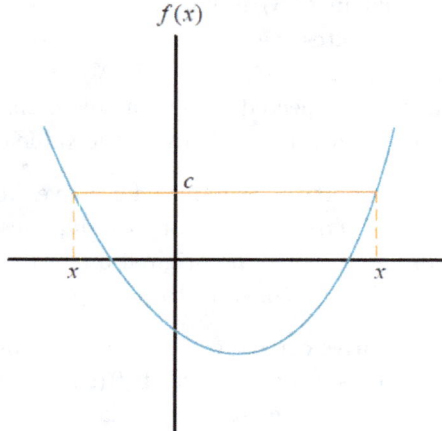

Figure 1.6. A closed interval.

The following is a well-known inequality between the arithmetic and the geometric mean.

> **Theorem**
>
> **Theorem 1.3.** *For any series of non-negative numbers x_i, $1 \leq i \leq n$,*
>
> $$\left(\prod_{i=1}^{n} x_i \right)^{\frac{1}{n}} \leq \frac{1}{n} \sum_{i=1}^{n} x_i, \qquad (1.4)$$
>
> *and in holds as an equality if and only if all $x_i, 1 \leq i \leq n$, are identical.*

Proof. Since the second derivative of $-\log x$ is $1/x^2$, we conclude that in the domain $x > 0$, $-\log x$ is a strictly convex function. Hence, for the choice of $\alpha_i = 1/n$, $1 \leq i \leq n$, by inequality (1.3), we get

$$-\log \frac{1}{n} \sum_{i=1}^{n} x_i \leq \frac{1}{n} \sum_{i=1}^{n} \log x_i.$$

Taking exponents on both sides concludes the proof of (1.4). Finally, by the definition of strong convexity, the inequality is strict in the case where all the entries are not equal. In the converse case where $x_i = c$, $1 \leq i \leq n$, for some constant c, it is easy to see that both sides of (1.4) equal to c. $\square$

Invoking (1.4) above, but now for $1/x_i$ instead of x_i, $1 \leq i \leq n$, leads to

$$\frac{n}{\sum_{i=1}^{n} \frac{1}{x_i}} \leq \left(\prod_{i=1}^{n} x_i \right)^{\frac{1}{n}}.$$

The left-hand side here is known as the *harmonic mean* of the series x_i, $1 \leq i \leq n$. In summary,

$$\frac{n}{\sum_{i=1}^{n} \frac{1}{x_i}} \leq \left(\prod_{i=1}^{n} x_i \right)^{\frac{1}{n}} \leq \frac{1}{n} \sum_{i=1}^{n} x_i. \qquad (1.5)$$

The fact that the harmonic mean is less than or equal to the arithmetic mean will be revisited in Section 3.2.3.

---

[a]This section is based on [9, p. 78].

1.3.2 Hölder and Cauchy–Schwarz inequalities

A straightforward generalization of (1.4) for the case where $n = 2$ is that for any two non-negative numbers x and y,

$$x^\alpha y^{1-\alpha} \leq \alpha x + (1 - \alpha)y, \quad 0 \leq \alpha \leq 1. \tag{1.6}$$

Assume now that a_i, b_i, $1 \leq i \leq n$, are two series each of which consists of n non-negative numbers. Also, suppose that $p > 1$ and $q > 1$ satisfy $1/p + 1/q = 1$. For some i, $1 \leq i \leq n$, take

$$x = \frac{a_i^p}{\sum_{j=1}^n a_j^p} \quad \text{and} \quad y = \frac{b_i^q}{\sum_{j=1}^n b_j^q}$$

and invoke it in (1.6) with $\alpha = 1/p$ and $1 - \alpha = 1/q$. We get that

$$\left(\frac{a_i^p}{\sum_{j=1}^n a_j^p}\right)^{\frac{1}{p}} \left(\frac{b_i^q}{\sum_{j=1}^n b_j^q}\right)^{\frac{1}{q}} \leq \frac{1}{p} \frac{a_i^p}{\sum_{j=1}^n a_j^p} + \frac{1}{q} \frac{b_i^q}{\sum_{j=1}^n b_j^q}.$$

Summing up the above with respect to i yields

$$\sum_{i=1}^n a_i b_i \leq \left(\sum_{i=1}^n a_i^p\right)^{\frac{1}{p}} \left(\sum_{i=1}^n b_i^q\right)^{\frac{1}{q}},$$

which is Hölder's inequality. For the case where $p = q = 2$, we get the well-known Cauchy–Schwarz inequality:

$$\sum_{i=1}^n a_i b_i \leq \sqrt{\sum_{i=1}^n a_i^2} \sqrt{\sum_{i=1}^n b_i^2}. \tag{1.7}$$

1.3.3 Chernoff bound

The following theorem states Chernoff's bound.[b]

---

[b]Based on [6, pp. 284–286].

> ### Theorem
>
> **Theorem 1.4.** *Let X be a random variable. Denote by $M_X(s)$ its moment-generating function, namely $M_X(s) = \mathrm{E}(e^{sX})$. Then, for any a,*
>
> $$P(X \geq a) \leq e^{-\phi(a)},$$
>
> *where*
>
> $$\phi(a) = \max_{s \geq 0}\{sa - \log M_X(s)\}.$$

Proof. Let $s \geq 0$ and define the random variable Y_a, which can only receive the following two values:

$$Y_a = \begin{cases} 0, & X < a, \\ e^{sa}, & X \geq a. \end{cases}$$

Clearly, $\mathrm{E}(Y_a) = e^{sa}P(X \geq a)$. Since $Y_a \leq e^{sX}$, we get that $\mathrm{E}(Y_a) \leq M_X(s)$. Combining all of the above, we can conclude that

$$P(X \geq a) \leq e^{-sa}M_X(s) = e^{-(sa - \log M_X(s))}.$$

Since the above holds for any s, it holds for its minimum value with respect to s. In other words,

$$P(X \geq a) \leq \min_{s \geq 0}\{e^{-(sa - \log M_X(s))}\} = e^{-\max_{s \geq 0}\{sa - \log M_X(s)\}},$$

as required. $\qquad\square$

Example (Standard normal distribution). In the case where X follows a standard normal distribution, $M_X(s) = e^{\frac{s^2}{2}}$. See, e.g., [19, p. 109]. Then, $\phi(a) = \max_{s \geq 0}\{sa - s^2/2\}$. It is easily seen that for $a \geq 0$, $\phi(a) = a^2/2$ as the maximization defining it is attained at $s = a$. Hence,

$$P(X \geq a) \leq e^{-\frac{a^2}{2}}, \quad a \geq 0.$$

As we see from the above example, Chernoff bound is useful in those cases where there is no explicit expression for the cumulative distribution function, as is the case with normally distributed random variables, while the corresponding moment generating function is readily accessible.

Exercise 1.2

Suppose that X follows a Poisson distribution with parameter λ, namely $P(X = i) = e^{-\lambda}\frac{\lambda^i}{i!}$ for any integer i, $i \geq 0$.

1. Show that $M_X(s) = e^{\lambda(e^s - 1)}$.
2. Use the Chernoff bound to show that for $a > \lambda$, $P(X \geq a) \leq e^{a-\lambda}(\frac{\lambda}{a})^a$.

1.3.4   *Maximum likelihood estimators (MLE)*

Maximum likelihood estimation is a commonly used technique for estimating parameters defining a distribution. Specifically, one first assumes that the distribution of a random variable belongs to some family of distributions, each of which is characterized by a specific value for a parameter (or parameters). In reality, this parameter (or parameters) is not known, but nevertheless one is interested in estimating it. To this end, a random sample from the underlying distribution is collected, and based on the observed realizations, one estimates the parameter under question. A number of estimation techniques exist, one of which is *maximum likelihood estimator*. The idea and the practice behind it are explained below in the following two examples.

Example 1 (Bernoulli distribution). Let X_i, $1 \leq i \leq n$, be n independent and identically distributed random variables, sometimes referred to as a *random sample*. Suppose that the common distribution of these random variables belongs to some family of distributions, each of which is characterized by a single-variable parameter. For example, suppose that $X_i \sim Ber(p)$, namely X_i can take either the value of 1 or 0 with probability p or $1-p$, respectively, $1 \leq i \leq n$, where $0 \leq p \leq 1$. Then, for any sequence of x_i with $x_i \in \{0, 1\}$, $1 \leq i \leq n$,

$$P(X_1 = x_1, \cdots, X_n = x_n) = \prod_{i=1}^{n} p^{x_i}(1-p)^{1-x_i} = p^{\sum_{i=1}^{n} x_i}(1-p)^{n-\sum_{i=1}^{n} x_i}.$$

$$(1.8)$$

In this setting, the focus is on the parameter defining the probability. This parameter is called the *likelihood function* and is denoted by $L(X_1, \ldots, X_n; p)$. The likelihood function is the same as the one stated in (1.8), except that the possible realizations x_i are replaced with X_i, $1 \leq i \leq n$, making these probabilities random variables themselves.

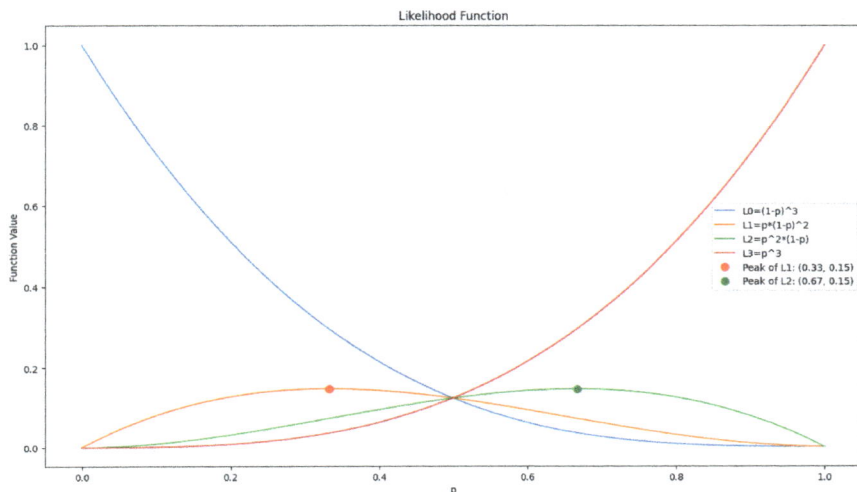

Figure 1.7. Likelihood functions.

See Figure 1.7 for the case where $n = 3$ and $\Sigma_{i=1}^{3} X_i$ can receive four possible values: $0, 1, 2,$ and 3. Then, the *maximum likelihood estimator* (MLE) for the parameter p is defined as the value for the parameter that maximizes the likelihood function. Note that this maximizing value is also a random variable. Thus, what we want to estimate is

$$\hat{p} \equiv \arg \max_{0 \leq p \leq 1} L(X_1, \ldots, X_n; p) = \arg \max_{0 \leq p \leq 1} p^{\sum_{i=1}^{n} X_i} (1 - p)^{n - \sum_{i=1}^{n} X_i}.$$

This function calls for optimization under the condition that $0 \leq p \leq 1$. Note that in the case where $\Sigma_{i=1}^{n} X_i = 0$ (respectively, $\Sigma_{i=1}^{n} X_i = n$), the likelihood function equals $(1 - p)^n$ (respectively, p^n), in which case it is easy to see that maximization is attained at $p = 0$ (respectively, $p = 1$). We next derive $\hat{p}$ for the general case. Specifically, instead of maximizing the likelihood function, we can maximize any other monotone increasing function of it. An intelligent choice is the log function. Then,

$$\log L(X_1, \ldots, X_n; p) = \log p^{\sum_{i=1}^{n} X_i} (1 - p)^{n - \sum_{i=1}^{n} X_i}$$

$$= \sum_{i=1}^{n} X_i \log p + \left(n - \sum_{i=1}^{n} X_i \right) \log(1 - p).$$

From Figure 1.8, we observe that the log-likelihood function receives only non-positive values. Note that the above equality is well defined

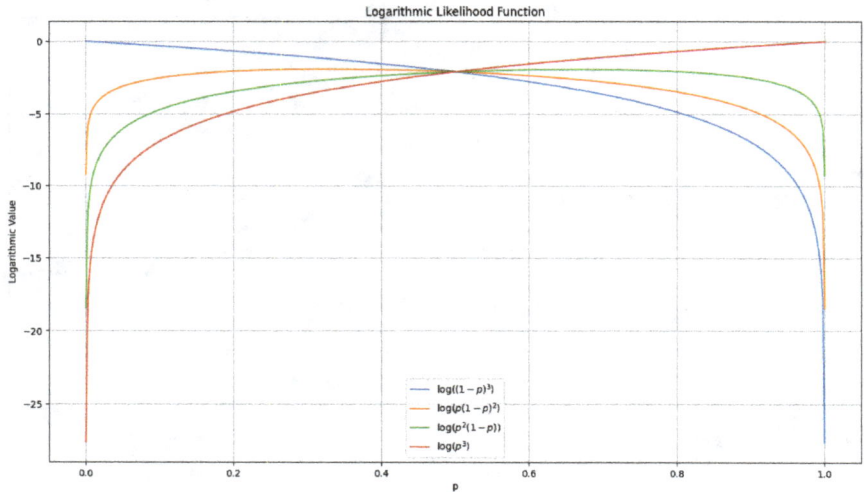

Figure 1.8. Log-likelihood functions.

for $0 < p < 1$. At the edges of the unit interval, we have the corresponding unbounded limits. Taking the derivative with respect to p, we get

$$\frac{\sum_{i=1}^{n} X_i}{p} - \frac{n - \sum_{i=1}^{n} X_i}{1 - p}.$$

Setting this derivative to zero, we get that the MLE for p is

$$\hat{p} = \frac{\sum_{i=1}^{n} X_i}{n},$$

namely the sample mean. Note that $\hat{p}$ is a fraction bounded between zero and one.[c] In particular, it is in the domain of the parameter p. Finally, we can check that the second derivative, which equals

$$-\frac{\sum_{i=1}^{n} X_i}{p^2} - \frac{n - \sum_{i=1}^{n} X_i}{(1 - p)^2},$$

is negative throughout the domain of parameter p, in which case the log-likelihood function is a concave function. In particular, the FOCs are sufficient in order to conclude that we have a global maximization.

---

[c]In the case where $\Sigma_{i=1}^{n} X_i = 0$ (respectively, $\Sigma_{i=1}^{n} X_i = n$), the limit of the log-likelihood function where $p \downarrow 0$ (respectively, $p \uparrow 1$) is zero, which is the largest possible value that the log-likelihood function can attain. Thus, MLE equals 0 (respectively, 1) in this case.

Example 2 (Poisson distribution). Suppose now that $X \sim Pois(\lambda)$ for some $\lambda \geq 0$. Then, the likelihood function is

$$\prod_{i=1}^{n} e^{-\lambda} \frac{\lambda^{X_i}}{X_i!} = e^{-n\lambda} \frac{\lambda^{\sum_{i=1}^{n} X_i}}{\prod_{i=1}^{n} X_i!}.$$

In the case where $\Sigma_{i=1}^{n} X_i = 0$, the function is $e^{-\lambda n}$, which is maximized over the domain $\lambda \geq 0$ at $\lambda = 0$. In general, the logarithm of the likelihood function equals

$$-n\lambda + \sum_{i=1}^{n} X_i \log \lambda,$$

up to an additive value that is not a function of lambda. Taking the derivative with respect to λ so that it gets the value of zero, we get that

$$-n + \frac{\sum_{i=1}^{n} X_i}{\lambda} = 0.$$

Thus, the MLE again equals the mean sample, that is $\frac{\sum_{i=1}^{n} X_i}{n}$, which is non-negative and is in the domain of the possible values for λ. The second derivative, $-\frac{\sum_{i=1}^{n} X_i}{\lambda}$, is non-positive and hence the log-likelihood is a concave function. In particular, the point derived is a global maximum point.

Exercise 1.3

A non-negative random variable is said to follow an exponential distribution with parameter λ, if its density function at point $x \geq 0$ equals $\lambda e^{-\lambda x}$.

1. Based on a random sample of such random variables, what is the likelihood function?
2. What is the MLE for λ?

1.4   Curve Fitting

The idea behind curve fitting for solving $\min_x f(x)$ is simple: Solving $\min_x f(x)$ is hard, so we define a simple function to be minimized instead; we call it $g(x)$, and solve $\min_x g(x)$ instead. For example, as we saw at the end of the previous section, if $g(x)$ is quadratic or cubic, solving $g(x)$ is simple. Solving $\min_x g(x)$ instead of $\min_x f(x)$ may remind us of the joke about a drunk person searching for his keys under the streetlight, instead of looking in the dark area where he lost them. Yet, if $g(x)$ is "close" enough to $f(x)$, then this is not a bad idea. Of course, it is not clear what we mean by "close" here.

Selecting a good $g(x)$ is called *curve fitting*. The standard approach is to fit an easy curve, i.e., a function that shares some local properties with $f(x)$ around some given point, say x_0. We next suggest four curve fitting methods.

(1) **Newton's fitting.** Look for a quadratic function $g(x) = \frac{1}{2}ax^2 - bx + c$, where $g(x_0) = f(x_0)$, $g'(x_0) = f'(x_0)$, and $g''(x_0) = f''(x_0)$. Clearly, $a = f''(x_0)$, $b = ax_0 - f'(x_0) = f''(x_0)x_0 - f'(x_0)$, and $c = f(x_0) + bx_0 - \frac{1}{2}ax_0^2$. Assuming that $a = f''(x_0) > 0$, then a minimum point for $g(x)$ is at $x = b/a$, which is easily seen to equal

$$x_0 - \frac{f'(x_0)}{f''(x_0)}. \tag{1.9}$$

See Figure 1.9.

(2) **The false position fitting.** A problem usually associated with Newton's curve fitting is the need to compute the second derivative at some point. If instead one has some knowledge on a point that is close to x_0, call it x_1, then $f''(x_0)$ can be replaced with the approximation to it, namely $\frac{f'(x_0) - f'(x_1)}{x_0 - x_1}$, which yields to

$$a = g''(x_0) = \frac{f'(x_0) - f'(x_1)}{x_0 - x_1},$$

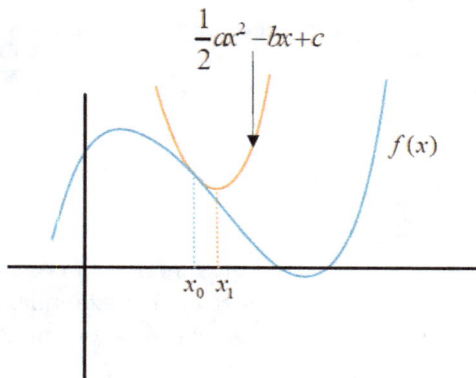

Figure 1.9. Newtons's curve fitting.

while the derivations for b and c are done in the same way as in Newton's method. The rationale behind this idea is the fact that

$$f''(x_0) = \lim_{x_1 \to x_0} \frac{f'(x_0) - f'(x_1)}{x_0 - x_1}.$$

(3) **The quadratic fitting.** Suppose that $(x_i, f(x_i))$ are given for $1 \leq i \leq 3$. There exists a unique quadratic function that crosses these three points. Denote this function by $g(x) = \frac{1}{2}ax^2 - bx + c$. Then a, b, and c are the unique solutions for the 3×3 linear equations:

$$\frac{1}{2}ax_i^2 - bx_i + c = f(x_i), \quad 1 \leq i \leq 3.$$

In order to find the value of x which minimizes $g(x)$, the aim is to find b/a. As it turns out, this value equals

$$\frac{b}{a} = \frac{1}{2} \frac{b_{23}f(x_1) + b_{31}f(x_2) + b_{12}f(x_3)}{a_{23}f(x_1) + a_{31}f(x_2) + a_{12}f(x_3)},$$

where $a_{ij} = x_i - x_j$ and $b_{ij} = x_i^2 - x_j^2$, $1 \leq i \neq j \leq 3$. See [18, p. 231] for further details.

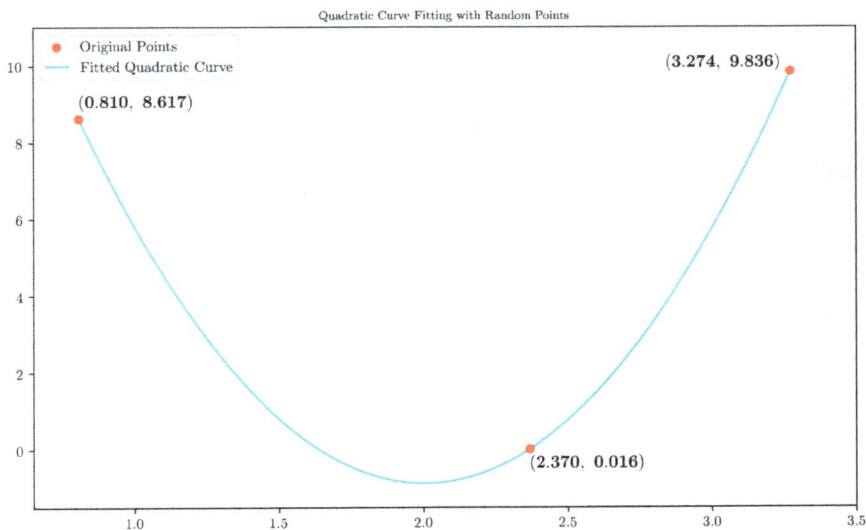

Figure 1.10. Quadratic curve fitting with three points: (1) $x_1 = 3.274$, $f(x_1) = 9.836$; (2) $x_2 = 0.810$, $f(x_2) = 8.617$; (3) $x_3 = 2.370$, $f(x_3) = 0.016$. The resulting parameters are: $a = 13.288$, $b = 26.647$, and $c = 25.850$.

Remark. It is well known that it is possible to construct an $n - 1$ polynomial that crosses n points, none of which share the same x value. In fact, there is a unique polynomial with this property. Thus, if $f(x)$ is known for n values of x, then one can approximate it (and then optimize the approximation) by the resulting $(n-1)$-degree polynomial. Specifically, if $f(x_i)$ are known for (different) x_i, $1 \leq i \leq n$, then the resulting polynomial is

$$g(x) = \sum_{i=1}^{n} f(x_i) \frac{\prod_{j \neq i}(x - x_i)}{\prod_{j \neq i}(x_j - x_i)}.$$

For further details, see, e.g., [14, pp. 116–117]. Of course, the quadratic curve fitting described above is just the special case where $n = 3$.

(4) **The cubic fitting.** Here too we assume that we have information on two points, x_0 and x_1. Specifically, suppose that the values of $f(x_0)$, $f(x_1)$, $f'(x_0)$, and $f'(x_1)$ are known. The idea is to construct a cubic function $g(x) = \frac{1}{3}ax^3 + \frac{1}{2}bx^2 + cx + d$ that shares with $f(x)$ these four

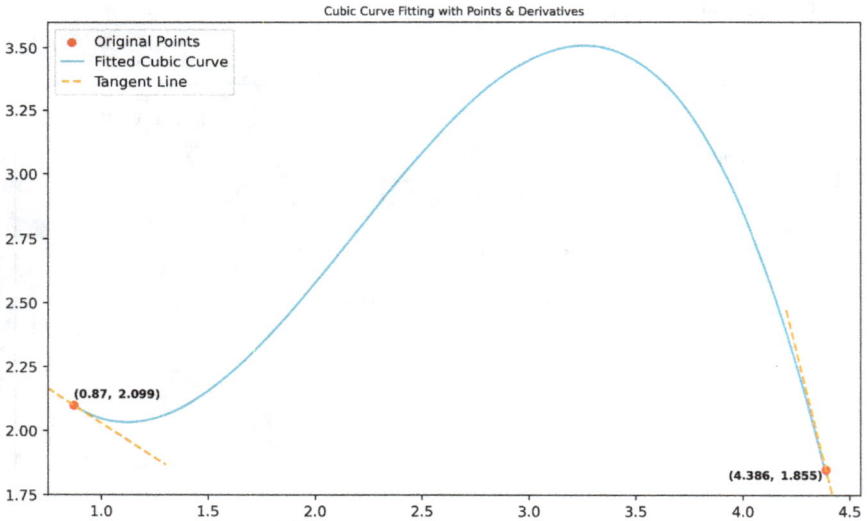

Figure 1.11. Cubic curve fitting with two points and their derivative values: (1) $x_1 = 0.870$, $f(x_1) = 2.099$, derivative value $f'(x_1) = -0.535$; (2) $x_2 = 4.386$, $f(x_2) = 1.855$, derivative value $f'(x_2) = -3.349$. The resulting parameters are $a = -0.909$, $b = 3.276$, $c = -3.310$, and $d = 3.673$.

values. Thus,

$$f(x_0) = \frac{1}{3}ax_0^3 + \frac{1}{2}bx_0^2 + cx_0 + d,$$

$$f(x_1) = \frac{1}{3}ax_1^3 + \frac{1}{2}bx_1^2 + cx_1 + d,$$

$$f'(x_0) = ax_0^2 + bx_0 + c,$$

and

$$f'(x_1) = ax_1^2 + bx_1 + c.$$

We get a 4×4 set of linear equations for the variables a, b, c, and d which can be solved by any linear equation solver. Once $g(x)$ is known, finding a point that meets the FOCs for $g(x)$ is straightforward: Since $g(x)$ is a cubic function, $g'(x)$ is quadratic and finding an x with $g'(x) = 0$ (if it exists) is easy. See Example 3 in Section 1.2.

1.5   Iterative Algorithms

With iterative algorithms, one generates a sequence of numbers in the hope that at least in the limit, they will converge to a local, if not global, minimum point. Another appealing feature might be that the corresponding functional values are non-increasing, if not strictly decreasing. How do such algorithms work? In general, a mapping from R to R is defined. Its input is the current (last generated) point (or the last few points in the same sequence) and its output is a new, hopefully, better approximation to the desired value. This procedure, with the new point as an input, is repeated (and repeated).

What would be a sensible mapping? This is where the curve fittings that were described in the previous section come in.

1.5.1   *Newton's method*

A good example is Newton's famous method. Specifically, denote by x_k the kth point generated by the procedure. Then define $g_k(x)$ in the same way suggested above around the point x_k, using the data $f(x_k)$, $f'(x_k)$, and

$f''(x_k)$. Then, define

$$x_{k+1} = \arg\min_x g_k(x).$$

In fact, as we can infer from (1.9),

$$x_{k+1} = x_k - \frac{f'(x_k)}{f''(x_k)}. \qquad (1.10)$$

The resulting algorithm is as follows:

Input: $x_0 \in R$, $\epsilon > 0$.

Output: an approximation to the minimal point.

Initialization: $k = 0$, $x_k = x_0$.

Iterative procedure:

$$x_{k+1} = x_k - \frac{f'(x_k)}{f''(x_k)}.$$

If $|x_{k+1} - x_k| < \epsilon$ STOP

$k \leftarrow k + 1$.

```python
# define the objective function and its first two derivatives
f = lambda x: x**3/3 - 4*x + 2
df = lambda x: x**2 - 4
ddf = lambda x: 2*x
# initialization point
x0 = 1.0

def newtons_method(f, df, ddf, x0, epsilon=1e-2, max_iterations=100):
    xn = x0
    iteration_values = []
    for n in range(max_iterations):
        # a loop to implement the iterative process
        fxn = f(xn)
        dfxn = df(xn)
        ddfxn = ddf(xn)
        # calculate the next x value using newton's formula
        next_xn = xn - dfxn / ddfxn
        iteration_values.append((n, xn, fxn, dfxn, ddfxn, next_xn))
        # check for convergence
        if abs(dfxn) < epsilon:
            break
        if ddfxn == 0:
            raise ValueError("zero derivative. no solution found.")
        xn = next_xn
    return iteration_values

iteration_stages = newtons_method(f, df, ddf, x0)

# formatting the iterations for display
iterations = [
    f"iteration {i+1}: x_{i} = {xn:.4f}, f(x_{i}) = {fxn:.4f}, f'(x_{i}) = {dfx:.4f}, f''(x_{i}) = {ddfx:.4f}, x_{i+1} = {next_x:.4f}"
    for i, xn, fxn, dfx, ddfx, next_x in iteration_stages]

x = sp.symbols('x')
display_f = x**3/3 - 4*x + 2

# print the function and the iterations
print(f"newton's method for the function: {display_f}
")

for iteration in iterations:
    print(iteration)
```

newton's method for the function: x**3/3 - 4*x + 2

iteration 1: x_0 = 1.0000, f(x_0) = -1.6667, f'(x_0) =-3.0000, f''(x_0) = 2.0000, x_1 = 2.5000
iteration 2: x_1 = 2.5000, f(x_1) = -2.7917, f'(x_1) =2.2500, f''(x_1) = 5.0000, x_2 = 2.0500
iteration 3: x_2 = 2.0500, f(x_2) = -3.3283, f'(x_2) =0.2025, f''(x_2) = 4.1000, x_3 = 2.0006
iteration 4: x_3 = 2.0006, f(x_3) = -3.3333, f'(x_3) =0.0024, f''(x_3) = 4.0012, x_4 = 2.0000

Figure 1.12. A Python implementation of Newton's method.

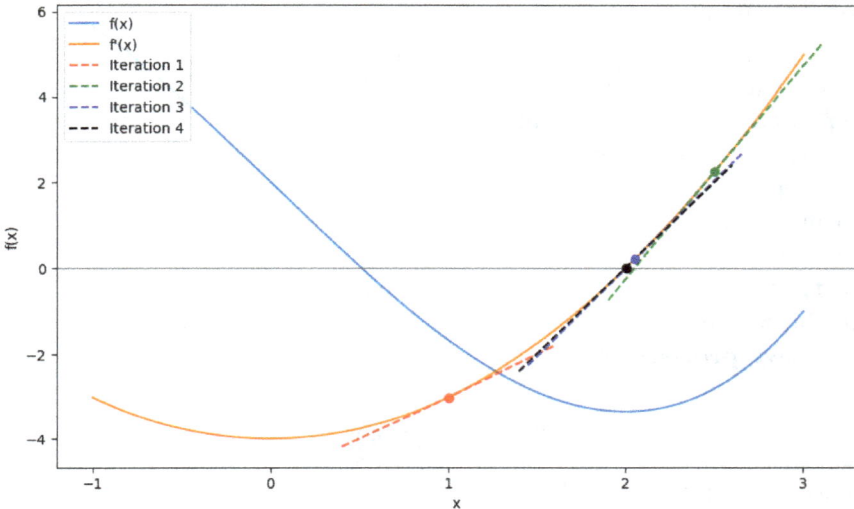

Figure 1.13. Newton's iterations.

We can say the following on this procedure. First, x_k is a fixed point of the mapping, namely $x_{k+1} = x_k$, if and only if x_k meets the FOCs for optimality. Second, if convergence is guaranteed to be reached for any $\epsilon > 0$, that is, if $\lim_{k \to \infty} x_k$ exists and it equals x^*, then x^* meets the FOCs. Yet, one cannot tell if the limit (if it exists) is a minimum, a maximum, or neither. By the same token, it is not guaranteed that the sequence $f(x_k)$, $k \geq 1$, is decreasing, and indeed examples where this is not the case exist. Finally, the value for $\epsilon > 0$ is referred to as the *tolerance level*.

Finally, we address the issue of local convergence. Suppose that x^* is a local minimum; does there exist a (small enough) $\epsilon > 0$, such that for any x_0 with $|x_0 - x^*| < \epsilon$, $\lim_{k \to \infty} x_k = x^*$ (i.e., the algorithm initialized at x_0 converges to x^*)? As we will show in the next section, the answer is "yes." We will also ask how fast this convergence is.

Remark. Note that if one is interested in finding the root of a function, that is, if one looks for y with $f(y) = 0$, then the iterative procedure is given by

$$x_{k+1} = x_k - \frac{f(x_k)}{f'(x_k)}.$$

1.5.2   *The false position method*

Recall that the false position curve fitting was similar to Newton's curve fitting with the exception that the second derivative of the focal point (in our case, the point x_k) is replaced by an approximation to it. As was said above, a possible approximation to it, once a sequence of points $x_1, x_2, \ldots$ is generated, is $\frac{f'(x_k) - f'(x_{k-1})}{x_k - x_{k-1}}$. The algorithm, which is also known as the secant method, now follows.

Input: $x_{-1}, x_0 \in R$, $\epsilon > 0$.

Output: an approximation to the minimum point.

Initialization: $k = 0$, $x_{k-1} = x_{-1}$, $x_k = x_0$.

Iterative procedure:

$$x_{k+1} = x_k - f'(x_k) \frac{x_k - x_{k-1}}{f'(x_k) - f'(x_{k-1})}.$$

If $|x_{k+1} - x_k| < \epsilon$ STOP

$k \leftarrow k + 1$.

Finally, what was said above on Newton's method in the last two paragraphs of the previous section holds verbatim here.

1.5.3   *The quadratic and cubic methods*

The quadratic and cubic methods are basically the same as the above two methods, and hence we will just highlight the cubic method here. Suppose that we have two current approximations to x^*, x_{k-1}, and x_k. Then, we can derive a cubic curve fitting function $g(x)$ based on $f(x_{k-1})$, $f(x_k)$, $f'(x_{k-1})$, and $f'(x_k)$. Then, x_{k+1} will be a point with $g'(x) = 0$. Recall that $g'(x)$ is a quadratic function, and therefore in the case where we have two roots, we should take the root with $g''(x) > 0$. This will be our x_{k+1}. See Example 3 in Section 1.2. The rest of the procedure is similar to the one described above.

1.6   Rate of Convergence

Suppose that a series a_n of real numbers converges to some limit, which without loss of generality is assumed to be zero. Consider now the series $b_n = |a_{n+1}/a_n|$. In the limit, we get something that is of the type $0/0$, which calls for inspection. We expect that the limit of this ratio (if it exists) to be less than 1. Looking at $\lim_{n \to \infty} b_n$, we can say, somewhat loosely, that the rate of convergence is faster in the case where the limit equals $1/3$ than in the case where the limit equals $1/2$. But is there a case where this limit equals 0? Here, we expect the rate of convergence to be the fastest. Indeed, now a_{n+1} is much closer to zero than a_n is, and therefore, even if we amplify a_{n+1} by dividing it by the almost zero value of a_n, we still get an almost zero value. Now, replace a_n in the denominator of b_n by a_n^p for some $p > 1$. In other words, the amplification done due to a division by a smaller number is now more dramatic. Do we still get a limit of zero? There is no clear answer here. Note that if p is too large, the limit will be infinity. Thus, very small values of p lead to a limit of zero, while very large values lead to a limit of ∞. Thus, we look for this knife-edge value where the limit is some number strictly between zero and infinity. This leads to the following definition.

Definition

Definition 1.4. The rate of convergence of the series a_n, $n \geq 1$, is the (not necessarily natural) real number p such that for some $0 < C < \infty$,

$$0 < \lim_{n \to \infty} \left| \frac{a_{n+1}}{a_n^p} \right| < C.$$

Note that the value of C is irrelevant here; although our instincts tell us that the smaller the (positive) value is, the faster the rate of convergence is.

Example. Let $a_n = 1/n^q$ for some $q > 0$. Then,

$$\frac{1/(n+1)^q}{1/n^q} = \frac{n^q}{(n+1)^q},$$

and its limit equals 1. Thus, we get a rate of convergence of 1, regardless of the value of q. This rate is called *a linear rate of convergence*. Slower rates are called *sublinear*, while faster are *superlinear*.

Theorem

Theorem 1.5. *Assume that $f''(x^*) \neq 0$. Then, the rate of convergence of the four iterative procedures suggested above are,*

- *Newton's*: 2,
- *false position*: 1.618,
- *quadratic fitting*: 1.3,
- *cubic fitting*: 2.

We next prove the first claim. In fact, we show that the rate of convergence in Newton's case is at least 2. As for the other three procedures, the reader is referred to [18, pp. 231–235].

Proof. For simplicity, denote $f'(x)$ by $g(x)$. By definition (see (1.10)) and the fact that $g(x^*) = 0$, we get that

$$x_{k+1} - x^* = x_k - x^* - \frac{g(x_k) - g(x^*)}{g'(x_k)}$$

$$= -\frac{g(x_k) - g(x^*) + g'(x_k)(x_k - x^*)}{g'(x_k)}.$$

Using Taylor expansion around the point x_k, we get that the above equals

$$\frac{g''(y)}{g'(x_k)}(x_k - x^*)^2,$$

for some y between x_k and x^*. Assume that x_k is close enough to x^* (and hence y is also close to x^*). Then, $|g''(y)|$ is bounded from above, say by C_1, and $|g'(x_k)|$ is bounded from below, say by C_2 (recall that we assume that $g'(x^*) \neq 0$). Hence,

$$|x_{k+1} - x^*| \leq \frac{C_1}{2C_2}|x_k - x^*|^2.$$

Since we can assume that $|x_k - x^*|\frac{C_1}{2C_2} < 1$, and hence $|x_{k+1} - x^*| < |x_k - x^*|$; the rest follows immediately. □

1.6.1 *Aitken acceleration*

Consider the following two approximations, which make sense when an iterative procedure gets closer to its limit x^*: For some q, $0 < q < 1$, representing the improvement at each step,

$$x_{n+1} - x^* \approx q(x_n - x^*),$$

and

$$x_{n+2} - x^* \approx q(x_{n+1} - x^*).$$

Then, treating these two approximations as a set of two linear equations with two unknown variables, q and x^*, and solving for x^*, we get, after some minimal algebra, that

$$x^* \approx x_{n+2} - \frac{(x_{n+2} - x_{n+1})^2}{x_{n+2} - 2x_{n+1} + x_n}. \tag{1.11}$$

Thus, instead of computing x_{n+3} via the original procedure, we can take the value on the right-hand side of (1.11) and make it as the x_{n+3}-th approximation point. This seems to be a good idea. The danger, however, is that now we can get carried away and compute x_{n+4} in the same way, using x_{n+1}, x_{n+2}, and x_{n+3} as input data, right? The problem is that no new functional values are involved, and we seem to be boot-strapping or fishing from the barrel and not from the sea. Thus, the idea behind this approach, known as Aitken acceleration, is to interfere with the original procedure periodically; say after every five or six iterations, conduct an Aitken acceleration step, go back to the original procedure, and so on.

1.7 Quasi–Convex Functions and Non-Derivative Based Procedures

Definition

Definition 1.5. Consider a function $f : [a, b] \to R$ for some finite a and b with $a < b$. We say that the function f is *unimodal, single-peak,* or *quasi-concave* in the interval $[a, b]$ if it has a unique local maximum point there. It is called *quasi-convex* if $-f(x)$ is quasi-concave.

Of course, an assumption of quasi-convexity along $[a, b]$ implies that the local minimum point is also the global minimum along this interval. Note that if this point, to be called x^*, is in the interior of $[a, b]$, namely it is attained at x^* with $a < x^* < b$, then the FOCs and the SOCs need to be met there. However, this is not necessarily the case if $x^* = a$ or $x^* = b$. Of course, an alternative definition for quasi-convexity is the existence of a point $x^* \in [a, \beta]$ such that the function is strictly decreasing along $[a, x^*)$ and is strictly increasing in $(x^*, b]$. Note that $x^* = a$ or $x^* = b$ is not ruled out. Finally, note that the key part of the definition of quasi-convexity is the uniqueness, since the existence of a global, and hence local, minimum point in any closed and bounded (i.e., compact) interval is a well-known fact.

A unimodal function: $f(x)$

x_0, where $f'(x_0) = 0$

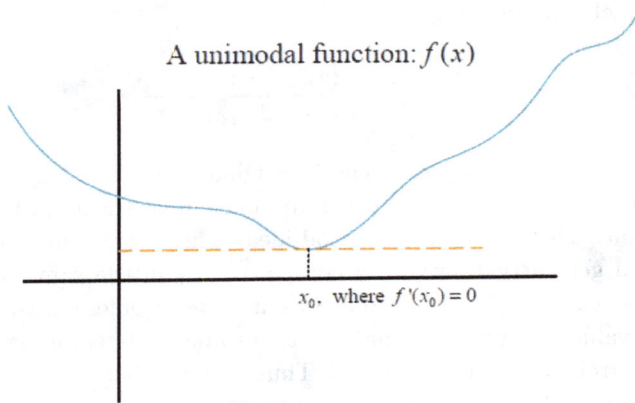

Figure 1.14. A quasi-convex function.

We next state two conditions that are met by such functions as well as by convex functions, as the latter group is a subclass of the former.

> **Theorem**
>
> **Theorem 1.6.** *For any quasi-convex function $f : [a, b] \to R$ we have:*
>
> 1. *For any $x, y \in [a, b]$ and any α with $0 \le \alpha \le 1$, $f(\alpha x + (1 - \alpha)y) \le \max\{f(x), f(y)\}$.*
> 2. *For any constant C, the set of points $\{x \mid f(x) \le C\}$ is an interval.*

Proof.

(1) First, note that for any x and y with $a \le x < y \le b$, the function is quasi-convex in $[x, y]$. For the proof itself, note the cases where $\alpha = 0$ and $\alpha = 1$ are trivial. By the way of contradiction, assume that for some $0 < \alpha < 1$, $f(\alpha x + (1 - \alpha)y) > f(x)$ and $f(\alpha x + (1 - \alpha)y) > f(y)$. Denote $\alpha x + (1 - \alpha)y$ by z. Consider the two compact non-empty intervals $[x, z]$ and $[z, y]$. A local minimum exists in each one of them. Moreover, these two minima are not the same point as neither of them can be the point z, the unique point that belongs to the two considered intervals. As constructed, each one of them is also a local minimum in

$[a, b]$. Yet, by definition, the function has only one local minimum in $[a, b]$. This is a contradiction.

(2) The proof basically repeats some of the arguments just used, and hence it is left for the reader.

$\square$

We can prove that the two properties stated above also hold for convex functions. The proof of the first property is trivial. As for the latter proof, see Item 2 of Theorem 1.2. Indeed,

Theorem

Theorem 1.7. *A strictly convex function is quasi-convex.*

The (easy) proof of this theorem is left as an exercise. Note that the converse statement is not true, even if "strictly convex" is replaced with the weaker requirement of "convex."

Suppose that one knows that x^*, the minimum point of $f(x)$ in the interval, lies in the sub-interval $[c, d]$, where $a \leq c \leq d \leq b$; then, this sub-interval is referred to as the *interval of uncertainty*. Clearly, $[a, b]$ is (the original) interval of uncertainty. Let $a = x_0 \leq x_1 \leq x_2 \cdots \leq x_n \leq x_{n+1} = b$ be a sequence of $n + 2$ points with $n \geq 1$. Note that the two boundaries of the original interval of uncertainty are included in this set of $n + 2$ points. Let

$$x_{k^*} = \arg \min_{0 \leq k \leq n+1} f(x_k).$$

Note that due to the quasi-convexity of the function, k^* can take at most two (consecutive) values. Assume first that k^* is unique (which is the more common case). Some thoughts (see Figure 1.15) will convince the reader that, firstly, if $k^* = 0$ or $k^* = n + 1$, then $x_k^* = x^*$ and, secondly, if $1 \leq k^* \leq n$, then $x_{k^*-1} \leq x^* \leq x_{k^*+1}$. This leads to a shrinkage of the interval of uncertainty from $[a, b]$ to $[x_{k^*-1}, x_{k^*+1}]$. In the case where there are two options for k^*, the uncertainty shrinks to the sub-interval they define. The question we pose here is what will be a good choice for these n points.

From now on we assume without loss of generality that $a = 0$ and $b = 1$.

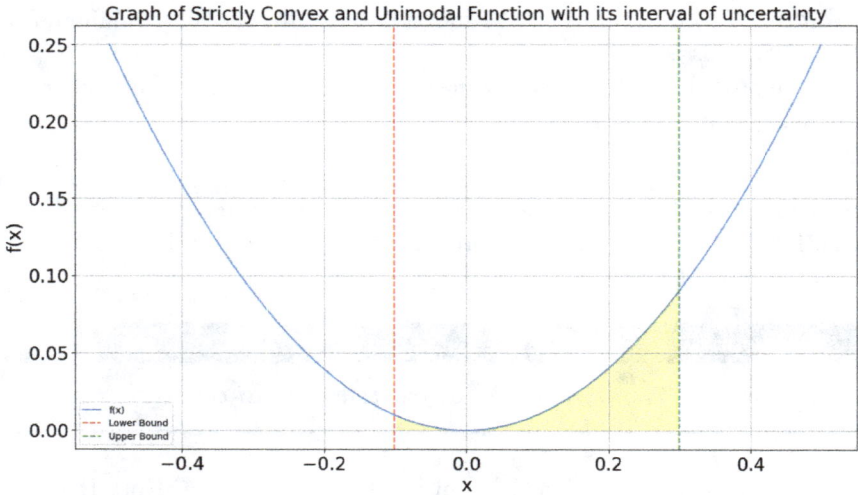

Figure 1.15. A strictly convex function with its interval of uncertainty.

1.7.1　*An offline search*

From the above discussion, we can infer that if one computes the value of a single point in the original uncertainty interval (in addition to the points 0 and 1), one cannot be certain of reducing this interval. For that, one needs to compute the value of the function at least at two points. Suppose that $n = 2$ and recall that $0 < x_1 < x_2 < 1$. Observe that the next interval of uncertainty will then be $[0, x_2]$ or $[x_1, 1]$. Our criterion for selecting the optimal x_1 and x_2 (under the constraint that $0 < x_1 < x_2 < 1$) is

$$\inf_{0 < x_1 < x_2 < 1} \max\{x_2 - 0, 1 - x_1\}.$$

In words, we want to minimize the largest possible width of the resulting uncertainty interval.

Our instinct might be to take $x_1 = 1/3$ and $x_2 = 2/3$, leading to $2/3$ being the width of the uncertainty interval. Yet, increasing x_1 toward (but not crossing) $1/2$, while decreasing x_2 toward (but not crossing) $1/2$, leads to a smaller interval. Indeed, the closer both x_1 and x_2 are to $1/2$ the better. However, making them both equal to $1/2$ brings us back to the uncertainty interval $[0, 1]$. The solution is to make $x_1 = 0.5 - \epsilon$ and $x_2 = 0.5 + \epsilon$ for some small $\epsilon > 0$. The width of the new interval will then be $0.5 + \epsilon$. To be sure, the smaller the value of ϵ, the better. We would like to note here that selecting x_1 and x_2 to be close to each other goes against our intuition in the search for optimal points since checking nearby values is usually less informative: The function is not expected to change by much.

Suppose now that $n = 3$. Now, the optimal choice under the above criterion is indeed the equal space choice, namely, $x_1 = 1/4$, $x_2 = 1/2$, $x_3 = 3/4$, leading to a slight improvement in the case where $n = 2$: Now, the width of the uncertainty interval equals exactly 0.5. Yet, the equal space results for any larger value of n, making $x_i = i/(n+1)$, $0 \leq i \leq n+1$, the optimal choice. The resulting width of the uncertainty interval is now $1/n$. Note that under any other choice, it is a matter of luck whether the uncertainty interval turns out to be shorter than $1/(n+1)$ but never longer. The choice we suggest here is optimal for the minimax criterion: In the worst case, we are guaranteed that the uncertainty interval is not longer than $1/(n+1)$.

This search for optimal points is an offline (or predetermined) search due to the fact that all points looked for are decided in advance. But perhaps an online search would be better. By that, we mean that for a given n, we evaluate the function f as some i points for some $1 \leq i < n$, so that the values of $f(x_j)$, $1 \leq j \leq i$, direct us to an intelligent choice for x_{i+1}. This is what we suggest in the next four subsections.

1.7.2 *Binary search*

For simplicity, assume that n is even and $f(0)$ and $f(1)$ are known. In the previous subsection, we learn that if two points are checked for their function value, the next uncertainty interval will have a width of $0.5 + \epsilon$. Moreover, we computed the function values at the edges of this interval. Thus, we can compute the function values of the two points that are $\pm\epsilon$ from the new midpoint. The new width will then be $(0.5 + \epsilon)/2 + \epsilon$. Repeating the computation will lead to an interval width of $0.25 + \epsilon/4 + \epsilon/2 + \epsilon$. It can thus be seen that $n/2$ repetitions (each consisting of two function evaluations) lead to an interval with width

$$(1/2)^{n/2} + 2(1 - 0.5^{n/2})\epsilon.$$

By ignoring the ϵ term here, we can see that for each computation of a function value, the uncertainty interval shrinks by a factor of $\sqrt{1/2} \approx 0.707$. As we will see below, the golden search yields better results. As we have already hinted above, the lack of efficiency of binary search is due to the fact that two nearby function values are computed at each iteration.

1.7.3   *The golden search*

Suppose that for some α, $0 < \alpha < 1/2$, two points are considered: α and $1 - \alpha$. The next uncertainty interval will be either $[0, 1 - \alpha]$ or $[\alpha, 1]$. In either case, we shall know the value function at the edges and at some interior point: In the former case, this interior point is α, while in the latter case it is $1 - \alpha$. A possible choice of α is one where α is relatively placed in the interval $[0, 1 - \alpha]$ as $1 - \alpha$ is placed in the original interval $[0, 1]$. In other words,

$$\frac{\alpha}{1 - \alpha} = \frac{1 - \alpha}{1}.$$

See Figure 1.16.

Solving for the shrinking factor $1-\alpha$, we get that $1-\alpha = \frac{-1+\sqrt{5}}{2} = 0.618$, which is the golden ratio. This is, of course, better than the binary search, which comes with a shrinking factor of 0.707. Is the golden ratio the optimal choice? We know that it is not in the case where $n = 2$ (where $\alpha = 0.5$ is optimal). Hence, it makes sense that the optimal choice depends on n. As we will see in the next subsection, the ith cut is a function of both i and n. Yet, as we will show, when $n = \infty$, this is the optimal cut throughout the entire procedure of searching for the optimal point.

1.7.4   *Fibonacci search*

First, a reminder. What is the Fibonacci sequence? Initializing with $F_0 = F_1 = 1$, define recursively, $F_N = F_{N-2} + F_{N-1}$, $N \geq 2$. This sequence turns out to be $1, 1, 2, 3, 5, 8, 13, 21, 34, 55, \ldots$

Second, the idea behind the algorithm we suggest here is that at each stage, we have an uncertainty interval with four known function values in it, two of which are at its edges. The two points in the middle cut the interval into three subintervals. This allows us to reduce the uncertainty interval by removing one of the extreme subintervals. In order to carry out the next cut, we now need to have four points in the other subinterval. As we already have three points, we need to look for the fourth point. If our goal is a minimax one, namely that we wish the worst (i.e.,

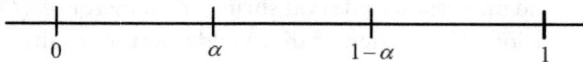

Figure 1.16. The golden search.

longest) next uncertainty interval to be as small as possible, the fourth point should be placed such that the distance between it and the edge closest to it is equal to the distance between the current middle point and the other edge.

Third, suppose that at stage k of the algorithm, the two middle points are located, relatively, at $1 - \alpha_k$ and α_k. Note that $\alpha_k > 0.5$, and this cut is in line with the previous paragraph: Both points are at a (relative) distance of $1 - \alpha_k$ from the closer, but opposite, edge. Next time, the cut will shrink the uncertainty interval by α_{k+1}. Suppose, without loss of generality, that the next subinterval is $[0, \alpha_k]$. Thus, looking ahead, we would like to know where the next (relative) α_{k+1} point is placed, namely we would like it to be at the previous $1 - \alpha_k$ point. This means that we require that

$$\alpha_{k+1} = \frac{1 - \alpha_k}{a_k}. \tag{1.12}$$

Note that this cut shrinks the uncertainty level by a factor of α_k.

Fourth, fix an integer $N \geq 3$ and let $F_1, F_2, \ldots, F_N$ be the first N Fibonacci numbers. Denote F_{N-k}/F_{N-k+1} by α_k, $1 \leq k \leq N - 2$. For example, $a_{N-2} = F_2/F_3 = 1/2$, $a_{N-3} = F_3/F_4 = 2/3$ and $\alpha_1 = F_{N-1}/F_N$. It is left as an exercise for the reader to show that this choice for a_k, $1 \leq k \leq N - 2$, meets (1.12). As said, at iteration k, the uncertainty interval shrinks by a factor of α_k. Thus, after $N - 2$ iterations, we get a shrinkage of

$$\prod_{k=1}^{N-2} \alpha_k = \prod_{k=1}^{N-2} \frac{F_{N-k}}{F_{N-k+1}} = \frac{1}{F_N}.$$

Reversing the point of view, if one wants the terminal uncertainty interval to be less than ϵ (the so-called tolerance level), then one should look for the minimum N with $(b_1 - a_1)/F_N \leq \epsilon$ and conduct N valuations of the function.

Last, suppose that one has a budget for N searches. The first two will be for the two edges of the original uncertainty interval. The question then is, what value for α_k, $1 \leq k \leq N-2$, is optimal — again optimal in the sense that it makes the worst possible uncertainty interval as short as possible. The answer is that it is the one suggested in the previous paragraph. The interested reader is referred to [17] for a proof. Thus, this choice leads to the so-called *Fibonacci search algorithm*. Of course, when N goes to infinity we are back to the golden search.

Remark. Since the final value α is 0.5, we get that at the final iteration the two mid-points, c and d, coincide. Thus, in order to proceed to the terminal iteration, one of the mid points needs to be perturbed a bit.

We next state the resulting procedure formally.

Fibonacci search algorithm
Input: $[a, b]$ (original uncertainty interval) and N (number of function valuations).
Output: the terminal uncertainty interval.
Initialization: $\alpha = F_{N-1}/F_N$, $c = a + (1 - \alpha)(b - a)$, and $d = a + \alpha(b - a)$.
Iterative part: For $k = 2$ until $k = N - 2$ do:
Let $\alpha = F_{N-k}/F_{N-k+1}$.
If $f(c) > f(d)$, then $a \leftarrow c$, $c \leftarrow d$ and $b \leftarrow b$ (the right subinterval), and $d \leftarrow a + \alpha(b - a)$ (next midpoint).
Otherwise, $a \leftarrow a$, $b \leftarrow d$ and $d \leftarrow c$ (the left subinterval), and $c \leftarrow a + (1 - \alpha)(b - a)$ (next midpoint).
$k \leftarrow k + 1$, REPEAT.
Print: "The terminal uncertainty interval is $[a, b]$ and the minimum computed functional value is $\min\{f(a), f(c), f(d), f(b)\}$."

A numerical example
Consider the function $f(x) = x^6 + 2x^5 - 3x^4 + 4x^3 - 5x^2 - 6x$ with an initial uncertainty interval $[0, 5]$. Note that $f(x)$ is quasi-convex in this interval. Suppose that one has a budget for $N = 6$ function valuations. Since $F_6 = 8$, this will lead to a terminal uncertainly interval with a width of $(5 - 0)/8 = 0.625$. We next give details on the procedure.

(1) The original value of α is $F_5/F_6 = 5/8 = 0.625$. The four points considered are $a = 0$, $c = 0 + (1 - 0.625)(5 - 0) = 1.875$, $d = 0 + 0.625(5 - 0) = 3.125$, and $b = 5$. The corresponding function values are $f(c) = f(1.875) = 50.261$ and $f(d) = f(3.125) = 1295.759$. Observe that $f(c) < f(d)$.

(2) Since $f(c) < f(d)$, the next uncertainty interval is the left subinterval, namely $[a, d] = [0, 3.125]$. Also, $\alpha = F_4/F_5 = 3/5 = 0.6$. Then, $a = 0$, $c = 0 + (1 - 0.6)(3.125 - 0) = 1.25$, $d = 1.875$, and $b = 3.125$. The new computed function value is $f(c) = f(1.25) = -4.906$. Observe that $f(c) < f(d)$.

(3) Since $f(c) < f(d)$, the next uncertainty interval is the left subinterval, namely $[a, d] = [0, 1.875]$. Also, $\alpha = F_3/F_4 = 2/3 = 0.666$. Then, $a = 0$, $c = 0 + (1 - 0.666)(1.875 - 0) = 0.625$, $d = 1.25$, and $b = 1.875$. The

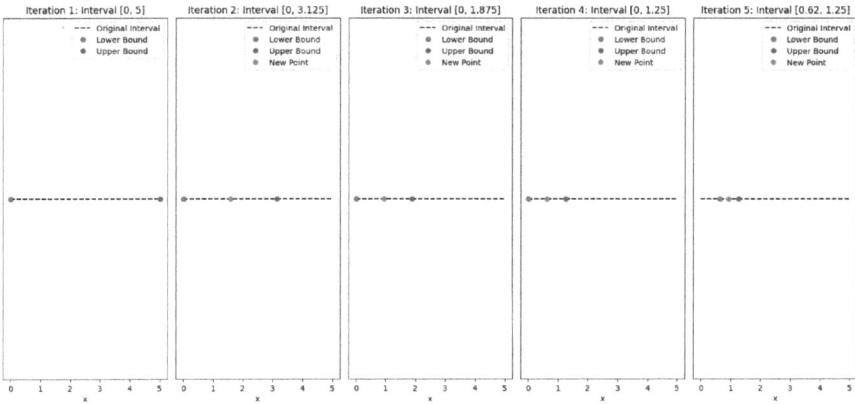

Figure 1.17. Five uncertainty intervals generated by the Fibonacci search algorithm.

new computed function value is $f(c) = f(0.625) = -4.934$. Observe that $f(c) < f(d)$.

(4) Since $f(c) < f(d)$, the next uncertainty interval is $[a, d] = [0, 1.25]$. Also, $\alpha = F_2/F_3 = 1/2 = 0.5$. Then, $a = 0$, $d = 0.675$, $c = 0 + (1 - 0.5)(1.25 - 0) = 0.625$, and $b = 1.25$. Note that the points c and d coincide, just as they should be in the final iteration. The new computed function value should be at a point close to this common value, for example, at 0.62, making $c = 0.62$ and $d = 0.65$. As $f(c) = f(0.62) = -4.982$, $f(d) = f(0.65) = -5.142$, and hence $f(c) > f(d)$, we conclude that the terminal uncertainty interval is $[c, b] = [0.62, 1.25]$. The lowest computed value is $\min\{f(c), f(d), f(b)\} = \min\{-4.982, -5.142, -4.906\} = -5.142$.

Chapter 2

Multi-Variable Unconstrained Optimization

2.1 Introduction

2.1.1 *Notation and preliminaries*

Single-variable functions were studied at length in the previous chapter. We now consider the more general case of multi-variable functions. Admittedly, from a technical point of view, the single-variable case is subsumed by the multivariable. Yet, we believe some intuition on single-variable functions is lost in the more general case.

Let $f(x) : R^n \to R$ be some function. Assume that it is differentiable as many times as needed. Denote by $\nabla f(x) \in R^n$ the vector with its n partial derivatives, called the *gradient* of the function at x, and by $\nabla^2 f(x) \in R^{n \times n}$ the symmetric matrix with $n \times n$ mixed partial derivatives, called the *Hessian matrix* of the function at point x. We use the notation $\nabla_y f(x)$ for the case where only the partial derivatives of a subset of variables, denoted here by y, are taken. The notation $\nabla^2_{zy} f(x)$ is used similarly. Of course, $\nabla^2_{zy} f(x) = (\nabla^2_{yz} f(x))^T$. Finally, for $x \in R^n$ we denote $\sqrt{\Sigma^n_{i=1} x_i^2}$ by $||x||$, which is referred to as the ℓ_2-norm of x.

The Taylor theorem tells us that the multi-variable version of (1.1) is

$$f(x) = f(x_0) + (x - x_0)^T \nabla f(x_0) + \frac{1}{2}(x - x_0)^T \nabla^2 f(y)(x - x_0), \quad (2.1)$$

for some $y \in R^n$ in the line connecting x_0 and x. Truncating the quadratic term in the Taylor expansion results in the function

$$f(x_0) + \nabla^T f(x_0)(x - x_0), \quad (2.2)$$

which is known as the *first-order* or *linear* approximation of the function $f(x)$ around the point x_0. In a similar fashion,

$$f(x_0) + \nabla^T f(x_0)(x - x_0) + \frac{1}{2}(x - x_0)\nabla^2 f(x_0)(x - x_0) \qquad (2.3)$$

is the *second-order* or *quadratic* approximation of the function $f(x)$ around the point x_0.

For a set of n functions $x_i(t) : R \to R$, $1 \leq i \leq n$, the function $x(t) = (x_1(t), \ldots, x_n(t)) : R \to R^n$ is called a *path*. We denote by $\dot{x}(t) \in R^n$ the vector $(\frac{dx_1(t)}{dt}, \ldots, \frac{dx_n(t)}{dt})$. Similarly, $\ddot{x}(t) = (\frac{d\dot{x}_1(t)}{dt}, \ldots, \frac{d\dot{x}_n(t)}{dt})$. The well-known chain rule says that

$$\frac{df(x(t))}{dt} = \sum_{i=1}^{n} \nabla f(x(t))_i \frac{dx_i(t)}{dt} = \nabla f(x(t))^T \dot{x}(t). \qquad (2.4)$$

Also, if $f(x) : R^m \to R^1$ and $A \in R^{m \times n}$, then for $g(x) = f(Ax)$ (and hence $g(x) : R^n \to R^1$), the gradient is

$$\nabla g(x) = A^T \nabla f(Ax). \qquad (2.5)$$

Finally, a symmetric matrix $A \in R^{n \times n}$ is said to be *non-negative* (respectively, *positive*) if for any $x \in R^n$ (respectively, any $x \in R^n$, $x \neq \underline{0}$) $x^T A x \geq 0$ (respectively, $x^T A x > 0$). Such matrices are known to have n real eigenvalues, and all of them are non-negative (respectively, positive). For details see, e.g., Chapter 10 in [14] or Example 6 in Section 4.2.1.[a]

2.1.2   *Convex sets and convex functions*

We briefly define convex functions. While the assumption that a function is convex makes our task much easier, it does not mean that we are looking for keys we lost in a dark alley under streetlight: In many applications, the assumption that a function is convex is not a severe limitation. Before getting to convex functions, we need the following definition:

Definition

Definition 2.1. A set $X \subseteq R^n$ is said to be convex if for any pair $x^1, x^2 \in X$, $\alpha x^1 + (1 - \alpha)x^2 \in X$ for any α with $0 \leq \alpha \leq 1$.

---

[a]In most texts the term "non-negativity" is replaced with the term "non-negative semi-definiteness", while "positivity" is replaced with "positive definiteness".

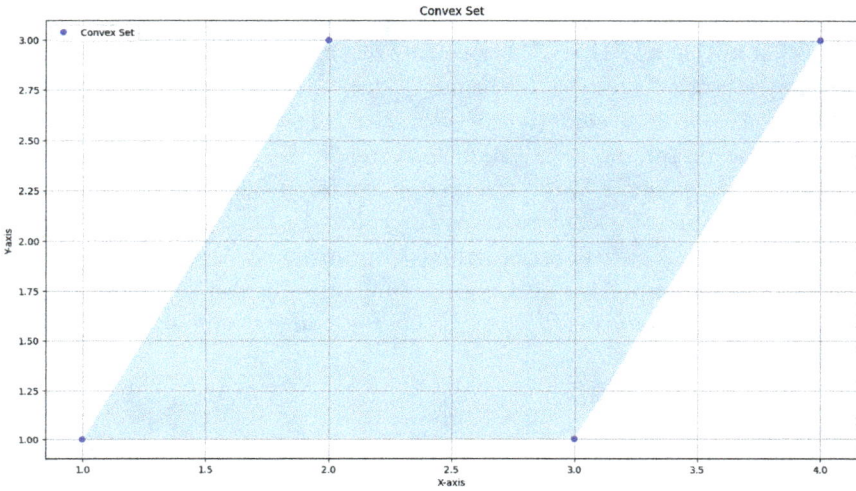

Figure 2.1. A convex set.

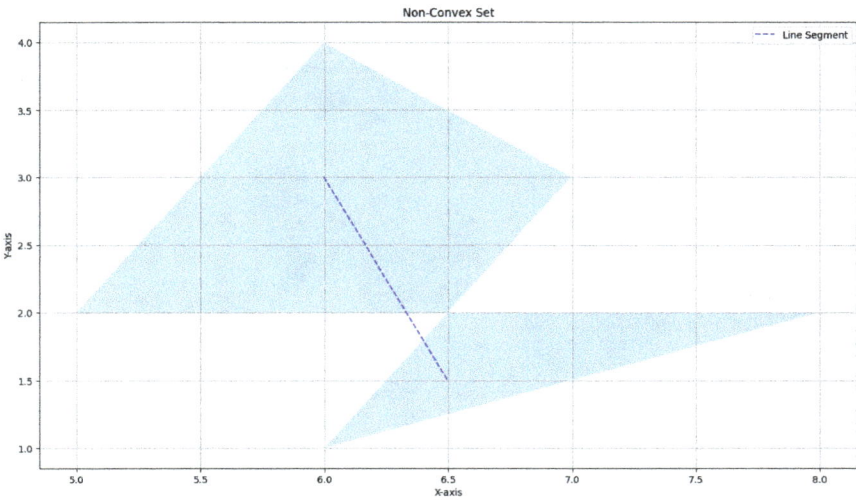

Figure 2.2. A non-convex set.

By an analogous argument to the one leading to (1.3), it is possible to show that in the case where X is a convex set, if $x^1, \ldots, x^k \in X$, and $\alpha_i \geq 0$, $1 \leq i \leq k$, with $\Sigma_{i=1}^k \alpha_i = 1$, then $\Sigma_{i=1}^k \alpha_i x_i \in X$. In this case, $\Sigma_{i=1}^k \alpha_i x_i$ is called a *convex combination* of $x^1, \ldots, x^k$. The empty set is an example of a convex set. So is any singleton. Yet, if the convex set X contains two different points x^1 and x^2, then it contains infinitely many points. A good example of a convex set is $X = R^n$. In fact, this is the default assumption on X when unconstrained optimization is defined. Another example of an empty set is a convex hull. Specifically, let $x^1, x^2, \ldots, x^k$ be a set of k points in R^n. Let C be the convex hull of these points, namely

$$C = \left\{ x \in R^n, \ x = \sum_{i=1}^k \alpha_i x_i \mid \sum_{i=1}^k \alpha_i = 1, \alpha_i \geq 0, 1 \leq i \leq k \right\},$$

where C is the set of all convex combinations of $x^1, x^2, \ldots, x^k$. It is easy to show that the convex hull of any given set of points is a convex set. We omit the details.

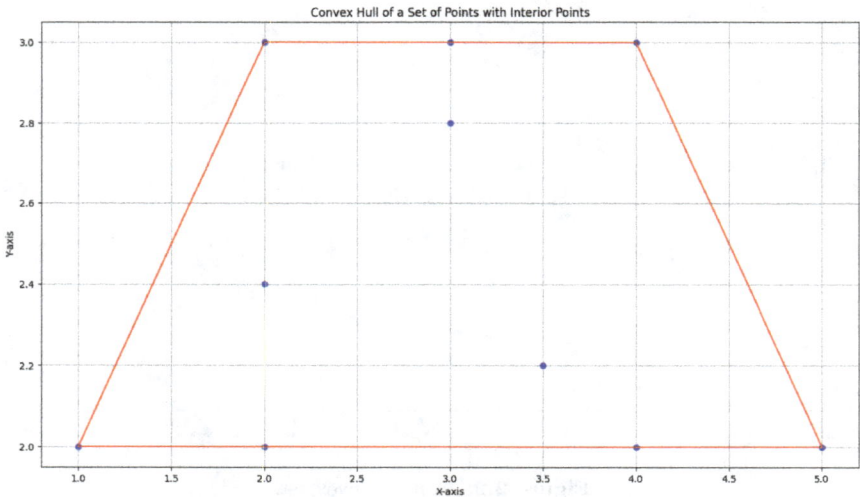

Figure 2.3. A convex hull.

Theorem

Theorem 2.1. *Let $f(x) : X \to R$, where X is an open convex set. Then, the following statements are equivalent:*

1. *For any pair $x^1, x^2 \in X$, and $0 \leq \alpha \leq 1$,*

$$f(\alpha x^1 + (1 - \alpha)x^2) \leq \alpha f(x^1) + (1 - \alpha)f(x^2). \qquad (2.6)$$

This condition can be expressed as "the function of the weighted average is less than or equal to the weighted average of the functions."

2. (*Gradient inequality*) *For any pair $x^1, x^2 \in X$,*

$$f(x^2) \geq f(x^1) + \nabla f(x^1)^T (x^2 - x^1). \qquad (2.7)$$

This condition can be expressed as "the function is greater than or equal to its tangent (or linear approximation) at any given point."

3. *For any $x \in X$, $\nabla^2 f(x)$ is non-negative.*

Proof.

- $(1) \Rightarrow (2)$. Let α satisfy $0 \leq \alpha \leq 1$. Then, by the convexity of f,

$$f(\alpha x^2 + (1 - \alpha)x^1) \leq \alpha f(x^2) + (1 - \alpha)f(x^1).$$

Equivalently,

$$\frac{f(x^1 + \alpha(x^2 - x^1)) - f(x^1)}{\alpha} \leq f(x^2) - f(x^1).$$

Taking limits when α goes to zero and using the chain rule (see (2.4)), we get from the above that

$$\nabla f(x^1)^T (x^2 - x^1) \leq f(x^2) - f(x^1),$$

as required.

- (2) $\Rightarrow$ (1). Let $x^1, x^2 \in X$, and $x = \alpha x^1 + (1 - \alpha)x^2$ for some α with $0 \leq \alpha \leq 1$. Clearly,

$$x^2 - x = -\frac{\alpha}{1 - \alpha}(x^1 - x). \tag{2.8}$$

Applying the gradient inequality on x and x^1, we get

$$f(x) + \nabla f(x)^T (x^1 - x) \leq f(x^1).$$

Applying the gradient inequality on x and x^2 invoking (2.8), we get

$$f(x) - \frac{\alpha}{1 - \alpha}\nabla f(x)^T (x^1 - x) \leq f(x^2).$$

Multiplying the first of the above two inequalities by α and the second by $1 - \alpha$, and then summing the two inequalities up, we get that

$$f(x) \leq \alpha f(x^1) + (1 - \alpha)f(x^2).$$

- (2) $\Rightarrow$ (3). Let $x \in X$ and let $d \in R^n$ be some direction. The assumption that X is an open convex set implies that for $\epsilon > 0$ small enough, $x + \epsilon d \in X$. By the gradient inequality (see (2.7)),

$$f(x + \epsilon d) \geq f(x) + \epsilon \nabla f(x)^T d.$$

Also, up to a term of the order of magnitude of ϵ^3,

$$f(x + \epsilon d) = f(x) + \epsilon \nabla f(x)^T d + \frac{\epsilon^2}{2} d^T \nabla^2 f(x)d.$$

Combining the above two identities, we conclude that the third term in the second expression is non-negative. The fact that this is true for any $d \in R^n$ completes the proof.

- (3) $\Rightarrow$ (2). Let $x^1, x^2 \in X$. Then there exists $x = \alpha x^1 + (1 - \alpha)x^2$ for some α with $0 \leq \alpha \leq 1$ (and hence $x \in X$) with the property that

$$f(x^2) = f(x^1) + \nabla f(x^1)^T (x^2 - x^1) + \frac{1}{2}(x^2 - x^1)^T \nabla^2 f(x)(x^2 - x^1).$$

The fact that the third term in the above identity is assumed to be non-negative completes the proof. $\qquad\square$

Definition

Definition 2.2. A function meeting one (and hence all) of the three conditions stated in Theorem 2.1 is called a convex function.

Convex function of two variables

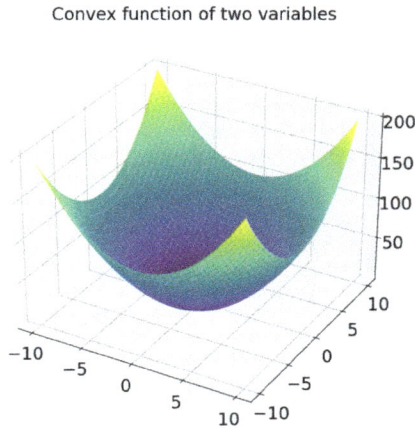

Figure 2.4. A convex function with two variables.

Example 1. For the matrix $A \in R^{m \times n}$, the function $f(x) : R^n \to R^m$, $f(x) = Ax$, is a convex function.

Exercise 2.1

Let $Q \in R^{n \times n}$ be a symmetric matrix and let (λ_i, ω_i), $1 \le i \le n$, be the set of pairs of eigenvalues and their corresponding eigenvectors.[b] Call vector $x \in R^n$ positively shaped if $x^T Q x \ge 0$. Define a negatively shaped vector accordingly.

(a) Show that ω_i is positively shaped if and only if $\lambda_i > 0$.
(b) Show that the set of positively shaped eigenvectors spans a linear subspace.
(c) Is the set of positively shaped vectors a convex set? Prove or construct a counterexample.

Definition

Definition 2.3. In the case where for all $x^2 \neq x^1$ and all α, $0 < \alpha < 1$, inequality (2.6) is strict, the function is said to be strictly convex.

---

[b]The existence of such eigensystems in the case of symmetric matrices is established in Example 6 in Section 4.2.1 below.

It is possible to see that in the case where $f(x)$ is strictly convex, inequality (2.7) is also strict (for $x^2 \neq x^1$ and $0 < \alpha < 1$). Likewise, $\nabla^2(x)$ is positive for $x \in X$.

Example 2. The quadratic function $f(x) : R^n \to R$, namely

$$f(x) = \frac{1}{2} x^T A x - b^T x + c,$$

is strictly convex for some symmetric matrix $A \in R^{n \times n}$, $b \in R^n$, and $c \in R$ if and only if A is positive. This is the case due to the fact that $\nabla^2 f(x) = A$ and by Item 3 of Theorem 2.1. A special case is where $A = I$, namely the identity matrix, and where $b = \underline{0}$. In this case $f(x) = \frac{1}{2}||x||^2$, and hence is a strictly convex function. Note that $||x||$ is also a convex function.[c]

Remark. Note that the above assumption that A is symmetric is without loss of generality. Specifically, if A is not symmetric, it can be replaced with the symmetric matrix $Q = \frac{1}{2}(A + A^T)$ as for any $x \in R^n$, $x^T Q x = x^T A x$. Also, in the case where Q is invertible, it is possible to assume that $b = \underline{0}$, as up to an additive vector that plays no role in optimizing a quadratic function, $(x - Q^{-1}b)^T Q(x - Q^{-1}b)$ and $x^T Q x - 2b^T x$ are equal. Indeed, $(x - Q^{-1}b)^T Q(x - Q^{-1}b) = x^T Q x - 2b^T x + b^T(Q_{-1})^2 b^T$, and the two functions agree up to a constant. Thus, if $y^* = \arg\min_y y^T Q y$, then the optimal solution is $x^* = y^* + Q^{-1}b$.

> **Definition**
>
> **Definition 2.4.** For a convex set X, $f(x) : X \to R$ is said to be (strictly) concave if $-f(x)$ is (strictly) convex.

The following theorem generalizes a similar result for the case where $n = 1$, and its proof is omitted as it repeats verbatim the proof given there.

---

[c]**Proof.**

$$||\alpha x^1 + (1 - \alpha)x^2|| \le ||\alpha x^1|| + ||(1 - \alpha)x^2|| = \alpha ||x^1|| + (1 - \alpha)||x^2||,$$

where the above inequality follows from the triangle inequality.

Theorem

Theorem 2.2. *A function* $f(x) : X \to R$ *is convex if and only if for any* $x^1, \ldots, x^m \in X$ *and any non-negative scalars* α_i, $1 \leq i \leq m$, *with* $\Sigma_{i=1}^m \alpha_i = 1$,

$$f\left(\sum_{i=1}^m \alpha_i x^i\right) \leq \sum_{i=1}^m \alpha_i f(x^i).^a$$

---

[a]The vector $\Sigma_{i=1}^m \alpha_i x^i$ is called a *convex combination* of x^i, $1 \leq i \leq m$.

The following theorem is stated without a proof. Proving it is left as an exercise for the reader.

Theorem

Theorem 2.3. *If* $f(x)$ *is a convex function, then for any* C *the set* $\{x | f(x) \leq C\}$ *is convex.*

Note that the above theorem is a necessary condition for a function to be convex, but it is not a sufficient one. Functions that meet this condition are called quasi-convex. They are well researched in the optimization literature (especially for functions $f(x) : R \to R$ like those described in the previous chapter), and we will not consider them any further in this text.

2.1.3 Goals and directions

The goal we introduce here is to find the minimum (or maximum) of a function $f(x) : R^n \to R$, called the *objective function* over all possible values for $x \in R^n$. The n variables associated with this function are sometimes refer to as the *decision variables*. As there are no restrictions on where in R^n the optimal point needs to be, the search for the optimal solution is called *unconstrained optimization*. The case where the search is limited to some subset of R^n is called *constrained optimization*. The latter is the subject of the second part of this book.

A few conclusions follow:

- **Optimality conditions.** As in the case where $n = 1$, one looks for conditions that are necessary and/or sufficient for a point to be optimal. A similar approach can be taken for the general case. These conditions might be helpful in refuting an optimality conjecture on a point that is not optimal, as well as in leading to algorithms for finding the optimal solution.

- **Finite vs. infinite algorithms.** One usually wishes to have an algorithm that takes as its input the function $f(x)$ and then yields a point $x^* \in \arg\min_{x \in R^n} f(x)$ as its output. As we have seen in the previous chapter where n was assumed to equal 1, this is usually too much to ask for. A modest requirement is that the algorithm generate a series of points with x^* as its limit. As an algorithm, it is assumed to be *iterative*, and the generated points are looked at as *approximations* to x^*. Again, having a finite convergence is too good to be true, and convergence is usually reached only at the limit, namely at infinity. Possessing this convergence property or not is, of course, not a question that the algorithm can answer. Hence, this issue needs to be resolved theoretically, i.e., in a mathematical way, normally prior to running the algorithm. Once the answer is positive, we can halt the algorithm somewhere in the process of generating points. This termination usually takes place once some *tolerance* criterion is met. By that, we mean that not much change has taken place during the last few iterations in the generated series of points and/or in the values of the objective function. The hope here is that if the tolerance criterion is met, then the current point is close enough to the optimal point. In general and a priori, there is no guarantee that this is the case. However, it might be the case for some specific algorithms. An alternative approach is to limit a priori the number of iterations in order to minimize the level of uncertainty. This was done above in the Fibonacci search and will be not considered again.
- **Local vs. global optimization.** Ideally, one is looking for $\arg\min_{x \in R^n} f(x)$. Yet, what we have learned from the previous section is that, given a point that is a candidate for being the optimal one, what one usually checks for is local optimization, namely whether the point is the best in the (possibly immediate) vicinity of the point. As the saying goes, "All politics is local."[d] The same is the case here, and unless some strong assumptions are imposed, one cannot deduce that a point that was discovered to be locally optimal is also globally optimal. The same can be said about convergence: At times convergence can be proved only for local optimization, not for global optimization.

---

[d]Attributed to Tip O'Neill, who was the speaker of the US House of Representatives in 1977–1987.

- **Minimization vs. maximization.** In multi-variable optimization, as in single-variable optimization, the corresponding FOCs are shared by both minimum and maximum points. Thus, meeting these conditions is a necessary condition for optimization, and once they are met, we cannot distinguish between a local minimum and maximum. Towards this end, we look for SOCs. Then, we can establish sufficient conditions, which, once coupled with the FOCs that are met, guarantee that the point is a local minimum or maximum. Of course, maximizing $f(x)$ is equivalent to minimizing $-f(x)$.

- **Computational complexity.** Each algorithm, given its input, has a running time. True, this is a vague statement as the running time depends on the code written and the computer used, but we can get around this objection by thinking of the running time as the number of simple algebraic calculations the algorithm has to perform, as a function of the size of the input (measured by how many numbers it contains), to get the desired result. Of course, we prefer an algorithm that requires the shortest possible running time. When we look at an iterative algorithm, namely one that generates a series of approximations to x^*, we want it to be one that requires less running time at each new iteration. However, this is not the only criterion. Another criterion is how much improvement in the approximation (called *rate of convergence*) we get at each iteration. Thus, there may be a trade-off between the two criteria. One might prefer an algorithm that converges slowly to one that does so faster, if the former significantly reduces the computational burden per iteration.

- **Line search.** In mathematics, we usually reduce a complicated problem to a simpler one that we already know how to solve. Unconstrained optimization is no exception: We will apply our skills in solving one-dimensional optimization problems to solving multidimensional ones. The approach is as follows. Let x be a point that is looked at as an approximation to x^*. Then the issue is in which *direction d* we should move away from x. Once the direction d is decided, we are back to a single-variable optimization problem. Specifically, we are looking for an optimal, or at least a good, value for t, called the *step size*, such that $x + td$ is a (hopefully substantial) improvement on x.[e] Finding such a t is a single-variable optimization problem. We have thus revealed the

---

[e]In machine learning circles, a step size is called the *learning rate*.

strategy we will adopt in the rest of this chapter: finding a good direction d (what we mean by "good" will be defined later). On the face of it, it seems that finding a good, not to say optimal, direction is as hard a task as finding the optimal x^*. Yet, as we will show, under two natural criteria to be defined later, the corresponding optimal ds are easily derived.

2.2 First-Order Conditions

Let $f(x) : R^n \to R$ be the function one wishes to optimize. Since we are dealing here with local conditions, we refer to the following definition.

> **Definition**
>
> **Definition 2.5.** A point $x^* \in R^n$ is said to be a local minimum for $f(x)$ if there exists some $\epsilon > 0$ such that for any $x \in R^n$ with $\|x - x^*\| < \epsilon$, $f(x^*) \leq f(x)$. Equivalently, for any $d \in R^n$, there exists some $\epsilon > 0$ such that $f(x^*) \leq f(x^* + \epsilon d)$. A similar condition defines the local maximum.

As in the single-variable case, a point x is said to be a *stationary point* if $\nabla f(x) = \underline{0}$.

Example. For the quadratic function $f(x) = \frac{1}{2}x^T Ax - b^T x + c$, $\nabla f(x) = Ax - b$. Thus, x^* is a stationary point if and only if $Ax^* = b$. Moreover, in the case where A is invertible, there exists a unique stationary point

$$x^* = A^{-1}b. \tag{2.9}$$

The corresponding objective function value equals

$$f(x^*) = -\frac{1}{2}b^T A^{-1}b + c.$$

In the following theorem we state the necessary *first-order condition* (FOC) for a point to be a local minimum or maximum.

> **Theorem**
>
> **Theorem 2.4.** *A necessary condition for a point x^* to be a local minimum or maximum is that it is a stationary point.*

Proof. Inspecting (2.1), it is always possible to find a point x close enough to x^* where the second term on the right-hand side dominates the third one. Moreover, if the point x^* is not a stationary one, it is possible to find x with either $(x - x^*)^T \nabla f(x^*) < 0$ or $-(x - x^*)^T \nabla f(x^*) > 0$, and hence x^* is neither a minimum nor a maximum point. $\qquad\square$

The condition $\nabla f(x^*) = \underline{0}$ is known as the *first-order condition* (FOC) for a point to be a local minimum. Note that it is the FOC described above for the case where $n = 1$, namely $f'(x^*) = 0$. Also, as exemplified by the cubic function in the single-variable case, this is a necessary condition but not a sufficient one.

Exercise 2.2

Consider the following function $f : R^2 \to R^{\mathrm{f}}$:

$$f(x, y) = \frac{x + y}{x^2 + y^2 + 1}.$$

(a) What is the gradient of this function?

(b) What are the two stationary points of f?

(c) What are the values of the function at the stationary points? Which one is a candidate for the global maximum and which one is a candidate for the global minimum?

(d) Show that for any (x, y), $f(x, y)$ is between the two values derived in the previous item. Hint: Argue that $f(x, y) \le \sqrt{2} \frac{\sqrt{x^2 + y^2}}{x^2 + y^2 + 1}$. Now use the fact that the right-hand side is a single-variable function (of $\sqrt{x^2 + y^2}$) whose global maximum equals $\sqrt{2}/2$.

(e) Prove that the two stationary points are indeed extreme points. Why was the previous item necessary?

---

[f]This equation appears in [3, p. 28].

2.3 Examples

2.3.1 *Maximum likelihood estimators for normal distributions*

Let $X \sim N(\mu, \sigma^2)$ and suppose that one looks for an MLE for the two parameters μ and σ^2 based on a random sample X_i, $1 \le i \le n$. Specifically, the likelihood function equals

$$\prod_{i=1}^{n} \frac{1}{\sqrt{2\pi}\sigma} e^{-\frac{(X_i-\mu)^2}{2\sigma^2}} = \left(\frac{1}{2\pi\sigma^2}\right)^{n/2} e^{-\frac{1}{2\sigma^2}\sum_{i=1}^{n}(X_i-\mu)^2}.$$

Its logarithm, if we ignore the additive terms that are not functions of μ and/or σ^2, equals

$$-\frac{n}{2}\log\sigma^2 - \frac{1}{2\sigma^2}\sum_{i=1}^{n}(X_i-\mu)^2.$$

Taking derivatives, once with respect to μ and once with respect to σ^2 (note that the derivative is not with respect to σ), we get

$$\frac{1}{\sigma^2}\sum_{i=1}^{n}(X_i-\mu) \tag{2.10}$$

and

$$-\frac{n}{2}\frac{1}{\sigma^2} + \frac{1}{2(\sigma^2)^2}\sum_{i=1}^{n}(X_i-\mu)^2. \tag{2.11}$$

The MLEs are derived by setting the above two expressions with zero. It is immediate that $\hat{\mu} = \overline{X}$, which is the sample mean, and that $\hat{\sigma^2} = \frac{1}{n}\sum_{i=1}^{n}(X_i-\hat{\mu})^2 = \frac{1}{n}\sum_{i=1}^{n}(X_i-\overline{X})^2$. Note that had σ^2 been given, the MLE for μ would have been the same. On the other hand, had μ been given, the MLE for σ^2 would have been $\frac{1}{n}\sum_{i=1}^{n}(X_i-\mu)^2$. In Section 2.4, it is proved that this stationary point is a local maximum point.

Exercise 2.3

Consider the function[g] $f : R^3 \to R$ defined as $f(x_1, x_2, x_3) = x_1(x_2 - 1) + x_3(x_3^2 - 3)$.

(a) Compute the gradient ∇f.

---

[g]This question is taken from [10, p. 114].

(b) Show that both $(0, 1, 1)^T$ and $(0, 1, -1)^T$ are stationary points of f. Are there any other stationary points?

(c) Show that none of the points derived in the previous item are local minima or maxima.

We emphasize that these are only necessary conditions and, even worse, that they are also necessary conditions for maxima. Thus, there is still some way to go before we reach our goal, but two points need to be made here: (i) Finding all points meeting the FOCs may reduce our search to a few points. In particular, once having a few such points may call for some ad-hoc techniques that select the local minimum among the candidates. (ii) As we will see later, there are special cases where these conditions are also sufficient. Spoiler alert, such cases occur when $f(x)$ is strictly convex.

2.3.2 *Maximum likelihood estimators for a logistic model*

Suppose that for some input vector $x \in R^m$ and for some parameters $a \in R^m$ and $b \in R$, the probability that an individual with input x possesses a given property equals

$$\frac{e^{a^T x + b}}{1 + e^{a^T x + b}}.$$

Suppose that n individuals are sampled, each with its own input. Denote the input of individual i by x_i, $1 \le i \le n$. Finally, let I_i be the (indicator) random variable that takes a value of 1 (respectively, 0) in the case where individual i possesses (respectively, does not possess) the property, $1 \le i \le n$. In particular,

$$P(I_i = 1) = \frac{e^{a^T x_i + b}}{1 + e^{a^T x_i + b}}, \quad 1 \le i \le n.$$

Of course, $P(I_i = 0) = 1 - P(I_i = 0)$, $1 \le i \le n$. As in (1.8), the likelihood function is then

$$L(I_i, 1 \le i \le n; a, b) = \prod_{i=1}^{n} \left(\frac{e^{a^T x_i + b}}{1 + e^{a^T x_i + b}} \right)^{I_i} \left(\frac{1}{1 + e^{a^T x_i + b}} \right)^{1 - I_i}.$$

Some algebra yields

$$\log L(I_i, 1 \le i \le n; a, b) = \sum_{i=1}^{n} I_i (a^T x_i + b) - \sum_{i=1}^{n} \log(1 + e^{a^T x_i + b}).$$

The first term is linear in all of the parameters, and in particular, it is a concave function. It is left to show that the second term is concave too. We leave the proof as an exercise for the reader. No explicit solution for the optimal values of the two parameters exists, calling for the use of iterative methods.

2.3.3 *Least squares solutions and regression*

The following problem is studied at length in [14], Section 7.1. Details on some of the concepts used in this section can be found there. Let $A \in R^{m \times n}$, where $m > n$. Also, assume that A is a full-rank matrix, namely $rank(A) = n$. In this case, the matrix $A^T A \in R^{m \times m}$ (which is clearly symmetric and positive) is invertible.[h]

For a given $b \in R^m$, the least squares problem is

$$\min_{x \in R^n} ||Ax - b||. \tag{2.12}$$

This optimization problem is sometimes referred to as an overdetermined system of linear equations. The reason behind this terminology is that under the condition stated here, the system of linear equations $Ax = b$ usually has no solution: There are too many equations (i.e., conditions to be met) and not enough variables (i.e., degrees of freedom). Thus, solving (2.12) is second best. Indeed, in real life, solving $Ax = b$ is too much to ask for, and the solution $\arg\min_{x \in R^n} ||Ax - b||$ is a good compromise.

Remark. Think of A_{ij} as the value of variable j observation i possesses, $1 \le i \le m$, $1 \le j \le n$. Likewise, b_i is the numerical value of variable $n + 1$ for individual i, $1 \le i \le m$. Thus, for a fixed j, all A_{ij}, $1 \le i \le m$, need to be measured using the same units of measurement; call them the units of column j, $1 \le j \le n$. Likewise, all b_i, $1 \le i \le m$, share the same units of measurement, call them the units of the right-hand side vector b. Since $A_{ij} x_j$ should be with the units of measurement as that of b_i, $1 \le i \le m$, $1 \le j \le n$ (as otherwise the expression $\Sigma_{j=1}^{n} A_{ij} x_j - b_i$ would be meaningless), we conclude that x_j is measured by the units of the vector b divided by the units of column j, $1 \le j \le n$.

$$\frac{1}{2}||Ax - b||^2 = \frac{1}{2}(Ax - b)^T(Ax - b) = \frac{1}{2}x^T A^T A x - b^T A x + \frac{1}{2}||b||^2.$$

---

[h]**Proof.** For $x \in R^n$, $A^T A x = \underline{0} \in R^m$ only if $x^T A^T A x = 0$, that is only if $||Ax|| = 0$. Since A is a full-rank matrix, this condition is met only when $x = \underline{0}$.

Taking the derivatives with respect to x and setting them with zero, we get that the FOCs are

$$A^T A x - A^T b = \underline{0}.$$

Observe that solving the least squares problem has been reduced to solving a problem of a system of linear equations with n equations and n decision variables. With the help of the full-rank assumption, we infer that they are solved by, and only by,

$$x^* = (A^T A)^{-1} A^T b. \tag{2.13}$$

Why this is a minimum point, and not, for example, a maximum point, is discussed in Section 2.4 below. Note that $A^T A$ (which is free of x) is the Hessian of the objective function, and it is also a positive matrix. The optimal objective value, as can be seen after some algebra, is equal to

$$\|A(A^T A)^{-1} A^T b - b\| = (b^T (I - A(A^T A)^{-1} A^T) b)^{\frac{1}{2}}. \tag{2.14}$$

The vector

$$A x^* = A(A^T A)^{-1} A^T b \tag{2.15}$$

is the closest vector to b out of all the vectors that are in the linear subspace spanned by the columns of A, and it is referred to as the *projection* of b on this subspace. Naturally, $b - A x^*$ is called the *residual*, while the optimal objective function value in (2.14) is the norm of the residual. This norm can be regarded as the distance between b and the above-mentioned linear subspace. It is also possible to see, with some algebra, that the projection and its residual are orthogonal, namely $(b - A x^*)^T A x^* = 0$. For more projections and residuals, see, e.g., Section 7.1 in [14]. In particular, see there the alternative proof of (2.13), which does not use derivatives.

Remark. The matrix $(A^T A)^{-1} A^T$ is called in the literature the *Moore-Penrose pseudo-inverse* of the matrix A, and it is usually denoted by $A^\dagger$:

$$(A^T A)^{-1} A^T. \tag{2.16}$$

This inverse exists for a full-rank tall matrix (namely a matrix for which the number of columns does not exceed the number of its rows). For more on this inverse see, e.g., Chapter 7 of [14].

An important application of the above analysis is linear regression. Suppose, as before, that m individuals are sampled for $n+1$ variables. Let A_{ij}

be the value of the jth variable for individual i, $1 \leq i \leq m$, $1 \leq j \leq n$, and let b_i be the corresponding value of the $n+1$-th variable. Then the entries of $x^* \in R^n$ are known as the coefficients of the linear regression of the $n+1$-th variable on the first n variables. The resulting linear function is useful in prediction. Specifically, if w_j, $1 \leq j \leq n$, are the values of the first n variables for a to-be-sampled individual, call it the $m+1$-th individual, then $\Sigma_{j=1}^{n} x_j^* w_j$ predicts (of course, with an error) the numerical value of the $n+1$-th variable of this individual.

The trivial case where $n = 1$ and $A_{i1} = 1$, $1 \leq i \leq m$ was already discussed in Example 2 in Section 1.2. In particular, it was shown there that $x^* = \frac{1}{m} \Sigma_{i=1}^{m} b_i$, which is the arithmetic mean of the entries of b. We next deal with another special case where $n = 2$.

Simple linear regression. There exists a special interest in the case where $n = 2$ and $A_{i1} = 1$, $1 \leq i \leq m$, known as *simple linear regression*. It is called so because only two variables are involved in this case; call them X and Y. Specifically, suppose that for m individuals, the values for these two variables are known. Denote them by (x_i, y_i), $1 \leq i \leq m$. One then looks for the line $y = q + px$, such that $\Sigma_{i=1}^{m}(y_i - q - px_i)^2$ is the lowest among all possible lines. Denote $\hat{y}_i = q + px_i$ as the *fitted value* and $y_i - \hat{y}_i$ as the *residual*, $1 \leq i \leq m$. We wish to minimize the sum of the squares of the residuals of the actual y values from the fitted ones. Thus, in this case,

$$A = \begin{pmatrix} 1 & x_1 \\ 1 & x_2 \\ \vdots & \vdots \\ 1 & x_m \end{pmatrix} \quad \text{and} \quad b = \begin{pmatrix} y_1 \\ y_2 \\ \vdots \\ y_m \end{pmatrix}.$$

Therefore,

$$A^T A = \begin{pmatrix} 1 & 1 & \cdots & 1 \\ x_1 & x_2 & \cdots & x_m \end{pmatrix} \begin{pmatrix} 1 & x_1 \\ 1 & x_2 \\ \vdots & \vdots \\ 1 & x_m \end{pmatrix} = \begin{pmatrix} m & \Sigma_{i=1}^{m} x_i \\ \Sigma_{i=1}^{m} x_i & \Sigma_{i=1}^{m} x_i^2 \end{pmatrix},$$

and

$$(A^T A)^{-1} = \frac{1}{m \Sigma_{i=1}^{m} x_i^2 - (\Sigma_{i=1}^{m} x_i)^2} \begin{pmatrix} \Sigma_{i=1}^{m} x_i^2 & -\Sigma_{i=1}^{m} x_i \\ -\Sigma_{i=1}^{m} x_i & m \end{pmatrix},$$

and

$$A^T b = \begin{pmatrix} 1 & 1 & \cdots & 1 \\ x_1 & x_2 & \cdots & x_m \end{pmatrix} \begin{pmatrix} y_1 \\ y_2 \\ \vdots \\ y_m \end{pmatrix} = \begin{pmatrix} \sum_{i=1}^m y_i \\ \sum_{i=1}^m x_i y_i \end{pmatrix}.$$

Finally,

$$\begin{pmatrix} q \\ p \end{pmatrix} = (A^T A)^{-1} A^T b$$

$$= \frac{1}{m \sum_{i=1}^m x_i^2 - (\sum_{i=1}^m x_i)^2} \begin{pmatrix} \sum_{i=1}^m x_i^2 & -\sum_{i=1}^m x_i \\ -\sum_{i=1}^m x_i & m \end{pmatrix} \begin{pmatrix} \sum_{i=1}^m y_i \\ \sum_{i=1}^m x_i y_i \end{pmatrix}.$$

Some additional algebra leads to the fact that the (simple) regression line of Y on X, $y = q + px$, is defined by the following two parameters: The slope

$$p = \frac{\sum_{i=1}^m y_i x_i - \frac{1}{m} \sum_{i=1}^m x_i \sum_{i=1}^m y_i}{\sum_{i=1}^m x_i^2 - \frac{1}{m}(\sum_{i=1}^m x_i)^2} = \frac{\text{Cov}(x,y)}{\text{Var}(x)}, \qquad (2.17)$$

and the intercept

$$q = \frac{1}{m} \sum_{i=1}^m y_i - \frac{p}{m} \sum_{i=1}^m x_i = \bar{y} - p\bar{x}, \qquad (2.18)$$

where $\bar{x} = \frac{1}{n}\sum_{i=1}^n x_i$, $\text{Cov}(x,y) = \frac{1}{n}\sum_{i=1}^n (x_i - \bar{x})(y_i - \bar{y})$, and $\text{Var}(x) = \text{Cov}(x,x)$. For further details see, e.g., [14], Section 7.2.

Remark. The case of simple linear regression can be proved in a much simpler way than is done above. In fact, the above proof was given in order to exemplify the general case. Specifically, recall that we are looking for

$$(p^*, q^*) = \arg\min_{p,q} \sum_{i=1}^m (y_i - q - px_i)^2.$$

Observe from (1.2) that for any fixed p, in particular, the optimal one, the corresponding best choice of q is $\overline{y - px}$, which equals $\bar{y} - p\bar{x}$. Hence, the problem we face is the single variable optimization problem of

$$\min_p \sum_{i=1}^m (y_i - \overline{y} - p(x_i - \overline{x}))^2.$$

This is a quadratic function in p:

$$\sum_{i=1}^m (y_i - \overline{y})^2 - 2p \sum_{i=1}^m (y_i - \overline{y})(x_i - \overline{x}) + p^2 \sum_{i=1}^m (x_i - \overline{x})^2,$$

where the minimization is attained at

$$p^* = \frac{\sum_{i=1}^m (y_i - \overline{y})(x_i - \overline{x})}{\sum_{i=1}^m (x_i - \overline{x})^2} = \frac{\operatorname{Cov}(x,y)}{\operatorname{Var}(x)}.$$

Finally, $q^* = \overline{y} - p^* \overline{x}$. The regression lines possesses a large number of nice properties. The interested reader is referred to [16].

Back to the general case. We have proved above that the solution is the unique stationary point of the function $||Ax - b||$. It is left to argue that this is the local minimum point, but this will be done in Section 2.6.1 below. Moreover, it will be shown there that this is also the global minimum.

Finally, one may wish to minimize the objective function

$$||Ax - b||^2 + \lambda ||x||^2, \tag{2.19}$$

for some constant $\lambda > 0$. For example, one may wish to minimize $||x||$ when λ is a penalty parameter. Put differently, the larger λ is, the more weight one puts on the second term in the objective function. Note that in order to make (2.19) meaningful and, moreover, in order to be able to assume that λ is a unit-free parameter, one needs to unify the units of the two terms defining the summation there. Since each of the columns of A may have different units of measurement, while making the x variables being in the corresponding units, the expression $||x||$ makes sense only in the case where each of the columns of A is first standardized before the ridge regression is applied.[i] Now the units of (2.19) coincide with the squares of the units of the b variable.

---

[i]By standardization, we mean that each entry in A is replaced by its original value minus the mean of its column, divided by the standard deviation of the entries in that column.

It is a simple exercise to show that the counterpart of (2.13) is now

$$x^* = (A^T A + \lambda I)^{-1} A^T b.$$

Note that the full-rank assumption with respect to A is not necessarily required here as long as $\lambda > 0$. This is known as *ridge regression*. In data science literature, other penalty functions are considered. Much interest exists in the case where the penalty $\lambda ||x||^2$ introduced in (2.19) above is replaced with $\lambda \Sigma_{i=1}^m |x_i|$, known as the LASSO (least absolute shrinkage and selection operator) regression.

2.3.4 *Logistic regression*

Looking again at the regression analysis done above, we can see that for a future sampled individual, we will have as input data values of variables similar to any of the rows of $A \in R^{n \times m}$ above, while using the regression line to predict its b value. But suppose now that the b variable is discrete or, to be more specific, is a binary yes or no variable. Now, the input on the right-hand side is, without loss of generality, $b_i \in \{-1, 1\}, 1 \leq i \leq m$. Thus, one possibility is to look whether $A_i x < 0$ or $A_i x > 0$ and determine the value of b_i, $1 \leq i \leq m$, accordingly. A possible criterion to judge whether or not x is a good solution is to look at $\Sigma_{i=1}^n b_i \operatorname{sgn}(A_i x)$. The larger its value, the better. Indeed, any time that the model is correct for a given individual i, we get $b_i \operatorname{sgn}(A_i x) = 1$. Otherwise, $b_i \operatorname{sgn}(A_i x) = -1$. Note that for any $x \in R^m$, $-n \leq \Sigma_{i=1}^n b_i \operatorname{sgn}(A_i x) \leq n$. This leads to a discrete optimization problem that is hard to solve and is also beyond the scope of this text. Another option is to optimize $f(\Sigma_{i=1}^n b_i A_i x)$ for some monotone increasing and continuous function $f : R^1 \to R^1$. Possible functions are e^x or $\log(1 + e^x)$. The latter choice is quite popular among practitioners and is called *logistic regression*. In summary, the optimization problem we face is

$$\max_{x \in R^m} \log(1 + e^{\Sigma_{i=1}^n b_i A_i x}).$$

Of course, the price one pays here is that the absolute values of $A_i x$, $1 \leq i \leq n$, play a role, although they should be irrelevant.

Once the optimal x is known and a new row, say A_{n+1}, is added, one can predict whether $b_{n+1} = 1$ or $b_{n+1} = -1$, depending on whether $A_{n+1} x$ is positive or negative, respectively. Alternatively, one can look at $1/(1 + e^{A_{n+1} x})$, which is a fraction between 0 and 1. The larger it is, the more likely it is that b_{n+1} is negative. By the same token, this ratio can be regarded as an estimation for the probability that b_{n+1} is negative.

The complementary value for 1 is, of course, one's estimation of the probability that it is positive. Note that the use of the terminology of probability here is somewhat informal as a probability model was not defined.

2.3.5 *Linear neural networks (LNN)*

Using the optimization problem posed in Section 2.3.3 on least squares as our point of departure, let now $X \in R^{m \times k}$ for some integer k (which does not necessarily equal 1 as was assumed there). Let $f : R^k \to R$ be some non-linear function. A possible model would say that up to some error term, b_i equals $f(A_i X)$, where A_i is the ith row of $A \in R^{n \times m}$, $1 \le i \le n$. Assuming that f is known, a possible optimization problem is to find the linear transformation matrix X that solves

$$\min_{X \in R^{m \times k}} \frac{1}{2} \sum_{i=1}^{n} (f(A_i X) - b_i)^2, \tag{2.20}$$

namely

$$X^* = \arg \min_{X \in R^{m \times k}} \frac{1}{2} \sum_{i=1}^{n} (f(A_i X) - b_i)^2. \tag{2.21}$$

In particular, we have an optimization problem with mk decision variables. Indeed, keeping in mind the regression approach, we can think of a pair of (A_i, b_i) as data that correspond to the ith sampled individual, $1 \le i \le n$. These n pieces of data are usually referred to as the *training set*. Then, while searching for the optimal X, one tries one's best to find fitted values $f(A_i X)$, $1 \le i \le n$, where the goal is to minimize the sum of the squares of the residuals.

In computer science, there is an interest in limiting the search for the linear transformation matrix X in the following way. For some integer $s \ge 1$ and some series of $s + 1$ integers $p_0, p_1, \ldots, p_s$ with $p_0 = m$ and $p_s = k$, define the s (decision) matrices $X_j \in R^{p_{j-1} \times p_j}$, $1 \le j \le s$. Each j here is usually referred to as a *layer*, $1 \le j \le s$, and naturally s is the number of layers. The optimization problem we face now is

$$\min_{X_j \in R^{p_{j-1} \times p_j}, 1 \le j \le s} \frac{1}{2} \sum_{i=1}^{n} \left(f \left(A_i \prod_{j=1}^{s} X_j \right) - b_i \right)^2. \tag{2.22}$$

Note that in searching for the optimal X_j, we optimize with respect to $p_{j-1} p_j$ decision variables, $1 \le j \le s$. This makes the total number of unknown variables equal[j] $\Sigma_{j=1}^{s} p_{j-1} p_j$. Note that AX is a linear function in all the decision variables, while any entry in $A_i \Pi_{j=1}^{s} X_j$ is a multinomial of degree s. Finally, let X_j^*, $1 \leq j \leq s$, be the optimal solution, namely

$$(X_1^*, \ldots, X_s^*) = \arg \min_{X_j \in R^{p_{j-1} \times p_j}, 1 \leq j \leq s} \frac{1}{2} \sum_{i=1}^{n} \left(f \left(A_i \prod_{j=1}^{s} X_j \right) - b_i \right)^2.$$

Consider now the gradient of the optimal objective value (2.22) with respect to the input row $A_i \in R^m$, $1 \leq i \leq n$. In this case, the decision variables are fixed to the values X_j^*, $1 \leq j \leq s$. Note that now, for an arbitrary input data $a \in R^m$, the function $f(a^T \prod_{j=1}^{s} X_j^*)$ is with $f : R^m \to R^1$. Its gradient is denoted by $\nabla_a f(a \prod_{j=1}^{s} X_j^*) \in R^m$. By the chain rule (see (2.5)), the gradient of the optimal objective value as a function of the input data A_i equals

$$\left(f \left(A_i \prod_{j=1}^{s} X_j^* \right) - b_i \right) \left(\prod_{j=1}^{s} X_j^* \right)^T \nabla_a f \left(A_i \prod_{j=1}^{s} X_j^* \right), \quad 1 \leq i \leq n.$$

Upon closer inspection, it can be seen that the above matrix-product $(\prod_{j=1}^{s} X_j^*)^T \in R^{k \times m}$ determines (locally, of course) the effect of an individual in the training set on the optimal objective value. This product equals $\prod_{j=s}^{1} (X_j^*)^T$, and it indicates how the effect of an individual propagates in the stated multiplicative way across the layers of the LNN.

Although technically this is not an issue, the problem posed above is appealing only in the case where $\Sigma_{j=1}^{s} p_{j-1} p_j < mk$, as still any solution of the latter version yields a solution to the former version but not the other way around. The model stemming from the limitation of the search for such s matrices is usually called a *linear neural network* (LNN). The idea behind LNN is that input vectors (which here are the rows of the matrix A) undergo, or percolate through, a series of linear transformations (whose number is denoted here by s), before undergoing a non-linear transformation (denoted here by f). These outputs need to be compared with the true observed values (denoted here by the corresponding entry in b). Following the least squares logic, we look for the optimal set of matrices,

---

[j]In fact, from this number we can subtract $s - 1$. This is the case since for each j, $2 \leq j \leq s$, one entry in X_j can be arbitrary fixed, without affecting the resulting optimal $\prod_{j=1}^{s} X_j$.

i.e., the optimal network. The choice of s, and likewise of the series of integers $p_0, \ldots, p_s$, calls for experience and is sometimes more of an art form than a science. The data, namely the matrix A and the right-hand side vector b, are called the training set, as usually the purpose of this exercise is to predict (of course, with an error) for an added row (or rows) of A, what should the corresponding right-hand side entry be. Alternatively, this training set of n vectors can be viewed as a sample from a larger population, and one wishes to extrapolate (again, with an error) from the training set to the rest of the population.

Remark. In LNN modeling, it is customary to consider $X_j + w_j$ (as opposed to X_j) where $w_j \in R^{p_j - 1}$ is an additional decision vector, $1 \leq j \leq s$. We assumed above that $w_j = \underline{0}$, $1 \leq j \leq n$, to simplify the exposition.

2.4   The Steepest Descent Method

In this subsection, we discuss a procedure for deriving the optimal point. Denote by x^* the optimal point one is after. Then, any point x can be looked at as an approximation to it, no matter how good or bad the approximation is. Next, we use x as a starting point (in fact, an input to some operation) and derive a better approximation for x^* (the output of this operation). Then, we use the output as the new input to the operation in each successive iteration, until we achieve a good approximation to x^*. The key challenge is how to design this procedure. Some suggestions are stated in this chapter.

Consider the affine function $x(t) : R \to R^n$, $x(t) = x + td$, for some point $x \in R^n$ and some direction $d \in R^n$. Note that $x = x(0)$. The function can be understood as a movement away from an initial point x in direction d, where t is the step size. This function is sometimes called a *path*. True, d and $2d$ move in the same direction, and so it makes sense to normalize d by assuming without loss of generality that $||d|| = 1$. Also, assume without loss of generality that $t \geq 0$. Otherwise, one could replace d with $-d$.

Consider the second Taylor polynomial expansion for $f(x(t))$ as a function of t around the point $t = 0$, where the third and higher terms are ignored:

$$f(x(0)) + t\nabla f(x(0))^T d + t^2 d^T \nabla^2 f(x(0))d$$
$$= f(x) + t\nabla f(x)^T d + t^2 d^T \nabla^2 f(x)d. \tag{2.23}$$

Assuming that $\nabla f(x)^T d \neq 0$, we can see that for $t > 0$ small enough, what determines whether or not $f(x(t)) < f(x)$, is the sign of $\nabla f(x)^T d$. Naturally, d is called a *descent direction* in x if $\nabla f(x)^T d < 0$. Define an *ascent direction* in a similar way, and note that if d is a descent direction then $-d$ is an ascent direction. Note also that the point of view here is local: The movement away from x is infinitesimal. Specifically, if d is a descent direction, then a small movement away from x in this direction guarantees an improvement. However, no such guarantee exists for a large movement away from x. Finally, note that once the direction is known, finding the optimal step size is a single-variable optimization problem.

There is an easy choice for a descent direction, namely $-\nabla f(x)$ or, after normalization, this is $-\nabla f(x)/||\nabla f(x)||$, which obviously has a norm of one. Indeed, $-\nabla f(x)^T \nabla f(x) = -||\nabla f(x)||^2 < 0$. More interestingly, this is the best choice in the following sense:

Theorem

Theorem 2.5.

$$-\nabla f(x)/||\nabla f(x)|| = \arg \min_{d \in R^n} \{\nabla f(x)^T d, \quad s.t. \quad ||d|| = 1\},$$

and

$$\nabla f(x)/||\nabla f(x)|| = \arg \max_{d \in R^n} \{\nabla f(x)^T d, \quad s.t. \quad ||d|| = 1\}.$$

Proof. The Cauchy–Schwarz inequality (see (1.7)) says that for any two vectors $a, b \in R^n$, $-||a||||b|| \leq a^T b \leq ||a||||b||$. Taking $a = \nabla f(x)$ and $b = d$ yields

$$-||\nabla f(x)||||d|| \leq \nabla f(x)^T d \leq ||\nabla f(x)||||d||,$$

which, due to the fact that $||d|| = 1$, yields

$$-||\nabla f(x)|| \leq \nabla f(x)^T d \leq ||\nabla f(x)||.$$

The choice $d = -\nabla f(x)/||\nabla f(x)||$ (respectively, $d = \nabla f(x)/||\nabla f(x)||$), makes the left (respectively, right) inequality binding. This completes the proof. $\square$

An alternative proof of this theorem is given as Example 10 in Section 4.2.1 below. In words, the theorem says that locally at x and given a constant infinitesimal step size, this is the best direction in the sense of improving the approximation to x^* per unit step. For obvious reasons, it is called the *steepest descent* direction. The next issue is what should the optimal step size be. Here, we do not restrict ourselves to small steps, and hence an underlying assumption here is that this optimization, known as a *line search* (where "line" here is a reference to "direction"), can be done in an efficient way. Needless to say, $f(x+t^*d) < f(x)$, where t^* is the optimal step size.

Thus, the steepest descent algorithm, sometimes referred as the gradient descent algorithm, is as follows:

Input: $x_0 \in R^n$, $\epsilon > 0$.

Output: An approximation to x^*.

Initialization: $i = 0$.

Iterations: Let $t_i^* = \arg\min_{t\geq 0} f(x_i - t\nabla f(x_i))$.
Let $x_{i+1} = x_i - t_i^*\nabla f(x_i)$.
If $||x_{i+1} - x_i|| \leq \epsilon$ STOP
$i \leftarrow i + 1$. REPEAT

In the above algorithm, we have suggested terminating the procedure when there is little change in the approximation in the current iteration. As with the single-variable optimization problem, another termination criterion is to stop when there is not much improvement in the corresponding objective function value, namely when $f(x_{i+1}) - f(x_i) \geq -\epsilon$. Finally, note the choice of the step sizes, t_i^*, $i \geq 0$. Yet, this choice may come with much effort per iteration. There are a number of other strategies for selecting the step size, but they are beyond the scope of this text. A possible source on this is [5, pp. 34–46].

The steepest descent algorithm converges to a local minimu point in the (typical) case where the sequence of generated approximations is bounded. See [5, p. 54], for a sufficient condition for the choice of a step size that guarantees convergence (under some mild conditions that the function needs to obey). The convergence is to a point where the FOCs are met, namely to a stationary point where $\nabla f(x) = \underline{0}$. Note that the set of points meeting the FOCs is the set of fixed points of the algorithm. In other words, $x_{i+1} = x_i$ if and only if $\nabla f(x_i) = \underline{0}$. Finally, note that the tolerance criterion $||x_{i+1} - x_i|| \leq \epsilon$ can be replaced with $||\nabla f(x_i)|| \leq \epsilon$ or any combination of the two versions.

```
 1  import numpy as np
 2  from scipy.optimize import minimize_scalar, approx_fprime
 3
 4  def f(x):
 5      # define a multivariate objective function
 6      return 0.5*(x[0] - 1)**2 + 0.3*(x[1] - 2)**2 + 0.2*(x[2] - 3)**2
 7
 8  def grad_f(x):
 9      # built-in function to calculate the approximate gradient for a given objective function
10      return approx_fprime(x, f, epsilon=1e-4)
11
12  def line_search(x, grad):
13      # line search for step size optimization - find optimal t in given range
14      func = lambda t: f(x - t * grad)
15      result = minimize_scalar(func, bounds=(0, 100), method='bounded')
16      return result.x if result.success else 0
17
18  def steepest_descent(f, grad_f, x0, max_iterations=100, epsilon=1e-3):
19      xi = x0
20      for i in range(max_iterations):
21          # a loop to implement the iterative process
22          f_val = f(xi)
23          grad = grad_f(xi)
24          t = line_search(xi, grad)
25          # calculate the next x value using the steepest descent formula
26          x_next = xi - t * grad
27
28          # print iteration details
29          print(f"iteration {i+1}: x_{i} = {xi.round(3)}, f(x_{i}) = {f_val.round(4)}, t = {t.round(3)}, "
30                f"grad_f(x_{i}) = {grad.round(3)}, x_{i+1} = {x_next.round(3)}\n")
31
32          # check for convergence
33          if np.linalg.norm(x_next - xi) < epsilon:
34              break
35          xi = x_next
36
37  # initial guess
38  x0 = np.array([0, 0, 0])
39  # run steepest descent optimization
40  steepest_descent(f, grad_f, x0)
```

Iteration 1: x_0 = [0 0 0], f(x_0) = 3.5, t = 1.59, grad_f(x_0) = [-1. -1.2 -1.2], x_1 = [1.59 1.908 1.908]

Iteration 2: x_1 = [1.59 1.908 1.908], f(x_1) = 0.4151, t = 1.271, grad_f(x_1) = [0.59 -0.055 -0.437], x_2 = [0.84 1.978 2.463]

Iteration 3: x_2 = [0.84 1.978 2.463], f(x_2) = 0.0706, t = 1.628, grad_f(x_2) = [-0.16 -0.013 -0.215], x_3 = [1.1 1.999 2.813]

Iteration 4: x_3 = [1.1 1.999 2.813], f(x_3) = 0.0121, t = 1.272, grad_f(x_3) = [0.101 -0. -0.075], x_4 = [0.973 2. 2.908]

Iteration 5: x_4 = [0.973 2. 2.908], f(x_4) = 0.0021, t = 1.631, grad_f(x_4) = [-0.027 -0. -0.037], x_5 = [1.017 2. 2.968]

Iteration 6: x_5 = [1.017 2. 2.968], f(x_5) = 0.0004, t = 1.268, grad_f(x_5) = [0.017 -0. -0.013], x_6 = [0.995 2. 2.984]

Iteration 7: x_6 = [0.995 2. 2.984], f(x_6) = 0.0001, t = 1.648, grad_f(x_6) = [-0.005 -0. -0.006], x_7 = [1.003 2. 2.995]

Iteration 8: x_7 = [1.003 2. 2.995], f(x_7) = 0.0, t = 1.245, grad_f(x_7) = [0.003 -0. -0.002], x_8 = [0.999 2. 2.997]

Iteration 9: x_8 = [0.999 2. 2.997], f(x_8) = 0.0, t = 1.751, grad_f(x_8) = [-0.001 -0. -0.001], x_9 = [1.001 2. 2.999]

Iteration 10: x_9 = [1.001 2. 2.999], f(x_9) = 0.0, t = 1.119, grad_f(x_9) = [0.001 0. -0.], x_10 = [1. 2. 3.]

Figure 2.5. A Python implementation of the steepest descent algorithm.

Exercise 2.4

Prove that if x_{k+1} is derived from x_k based on the exact line search when the steepest descent method is applied, then the gradients of f at x_k and at x_{k+1} are orthogonal.

The result of this exercise implies that in the case where $n = 2$, all odd directions coincide, and likewise all even directions coincide.

This phenomenon is called zigzagging. Perhaps taking a "shortcut" and moving along in between directions will lead to a faster procedure.

2.4.1 *Rate of convergence for quadratic functions*

Let $f(x) = \frac{1}{2}x^T Q x - b^T x + c$ for a symmetric positive matrix $Q \in R^{n \times n}$, a vector $b \in R^n$, and a constant $c \in R$. Since we are interested in $\arg\min_x f(x)$, we will assume throughout that $c = 0$. Clearly, $\nabla f(x) = Qx - b$ and $\nabla^2 f(x) = Q$. Moreover, $\nabla f(x) = \underline{0}$ if and only if $x = Q^{-1}b$. Recall that as Q is positive, it is also invertible.

Exercise 2.5

Let $A \in R^{m \times n}$, where $m > n$. Assume that it is a full-rank matrix.

(a) What is the rank of A?
(b) Prove that both $A^T A$ and $A A^T$ are symmetric.
(c) What is the rank of each of the above matrices? Which one, if any, is invertible?

As we have seen above, optimizing a quadratic function is as hard as inverting a square matrix. Nevertheless, let us see what happens when the steepest descent algorithm is applied to a quadratic function. First, note that we replace an $O(n^3)$ (one-time) procedure of a matrix inversion with an iterative algorithm that requires $O(n^2)$ operations per iteration (due to the required computation of $\nabla f(x_i) = Qx_i - b$). For large values of n, this might not be a bad idea. Second, closer to the optimal solution, dealing with a quadratic approximation for any function (using Taylor's series expansion) may result in some qualitative observations that shed some light on the original function. In other words, in our discussion, one needs to have in mind that the matrix Q here reflects $\nabla^2 f(x_i)$, which in turn approximates $\nabla^2 f(x^*)$.

For ease of notation, let $g(x) = Qx - b$, which is the gradient in this specific case. The key feature here is that the line search leads to a simple procedure as the single-variable search boils down to solving a single-variable quadratic function. We next give some details. Specifically, for the scalar t, let

$$h(t) = f(x - tg(x)) = \frac{1}{2}(x - tg(x))^T Q(x - tg(x)) - b^T(x - tg(x)).$$

This a quadratic function in t. It is easy to see that the quadratic coefficient equals $\frac{1}{2}g(x)^T Q g(x)$, while the linear coefficient equals

$$-g(x)^T Q x + b^T g(x) = g(x)^T (-Qx + b) = -g(x)^T g(x) = -||g(x)||^2.$$

Note that the free coefficient is irrelevant as far as the minimization is concerned. Thus, the optimal point t equals

$$t^* = \frac{||g(x)||^2}{g(x)^T Q g(x)}. \tag{2.24}$$

There is a practical lesson to be gained from the above. Inspect (2.23). It gives a quadratic approximation to the function $f(x)$, where Q is the Hessian matrix $\nabla^2 f(x_i)$ and $b = \nabla f(x_i)$. In particular, the gradient of this quadratic function equals $\nabla^2 f(x_i)x_i - \nabla f(x_i)$. Thus, a seemingly promising step size in the direction $-\nabla f(x_i)$ is

$$\frac{||\nabla^2 f(x_i)x_i - \nabla f(x_i)||^2}{(\nabla^2 f(x_i)x_i - \nabla f(x_i))^T \nabla^2 f(x_i)(\nabla^2 f(x_i)x_i - \nabla f(x_i))}.$$

We, later in Theorem 2.6, argue for this choice, but first we need to state the following inequality, called the Kontorovitch inequality.

Lemma

Lemma 2.1. *Let λ_1 and λ_n be the smallest and largest eigenvalues of a positive square matrix Q, respectively.[a] Then, for any $x \in R^n$,*

$$\frac{||x||^4}{(x^T Q x)(x^T Q^{-1} x)} \geq \frac{4\lambda_1 \lambda_n}{(\lambda_1 + \lambda_n)^2}. \tag{2.25}$$

---

[a]Both are known to be real and positive. The existence of such eigenvalues in the case of symmetric matrices is established in Example 6 in Section 4.2.1 below. See also [14, p. 139].

Proof. Let (λ_i, v_i), $1 \leq i \leq n$, be the n eigenpairs of the matrix Q, where the eigenvectors are known to be orthonormal. See [14, p. 105]. Let $x \in R^n$. Then, for some α_i, $1 \leq i \leq n$, $x = \Sigma_{i=1}^n \alpha_i v_i$. Also, $||x||^2 = \Sigma_{i=1}^n \alpha_i^2$. Clearly, $Qx = \Sigma_{i=1}^n \alpha_i \lambda_i v_i$ and hence $x^T Q x = \Sigma_{i=1}^n \alpha_i^2 \lambda_i$. Since $(1/\lambda_i, v_i)$, $1 \leq i \leq n$, are the eigenpairs of Q^{-1}, we infer that $x^T Q^{-1} x = \Sigma_{i=1}^n \alpha_i^2/\lambda_i$.

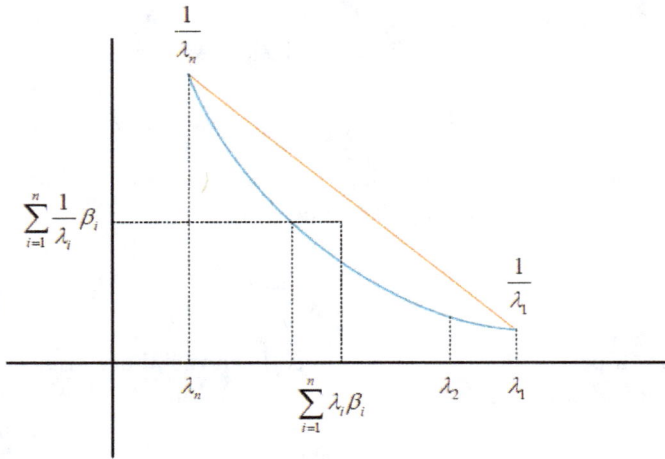

Figure 2.6. Kontorovitch inequality.

Hence,

$$\frac{\|x\|^4}{(x^T Q x)(x^T Q^{-1} x)} = \frac{(\sum_{i=1}^n \alpha_i^2)^2}{\sum_{i=1}^n \alpha_i^2 \lambda_i \sum_{i=1}^n \alpha_i^2 / \lambda_i}. \tag{2.26}$$

Denote $\alpha_i^2 / \Sigma_{j=1}^n \alpha_j^2$ by β_i, $1 \le i \le n$. Clearly, $\beta_i \ge 0$, $1 \le i \le n$, and $\Sigma_{i=1}^n \beta_i = 1$. Hence, the eigenpairs can be viewed as weights. Dividing both the numerator and denominator of (2.26) by $(\Sigma_{i=1}^n \alpha_i^2)^2$, we get

$$\frac{\frac{1}{\sum_{i=1}^n \lambda_i \beta_i}}{\sum_{i=1}^n \frac{1}{\lambda_i} \beta_i}.$$

The numerator is in fact the inverse of some weighted average between λ_1 and λ_n, say $\gamma \lambda_1 + (1 - \gamma)\lambda_n$, for some $0 \le \gamma \le 1$. We can also see that the denominator of the function $y = 1/x$ is greater than $\gamma/\lambda_1 + (1-\gamma)/\lambda_n$. Thus, this ratio is less than

$$\frac{\frac{1}{\gamma \lambda_1 + (1-\gamma)\lambda_n}}{\frac{\gamma}{\lambda_1} + \frac{1-\gamma}{\lambda_n}},$$

which, with some simple algebra, is shown to equal

$$\frac{\lambda_1 \lambda_n}{(\gamma(\lambda_n - \lambda_1) + \lambda_1)(\gamma(\lambda_1 - \lambda_n) + \lambda_n)}. \tag{2.27}$$

Minimizing the above with respect to γ is equivalent to maximizing the quadratic function in γ, $(\gamma(\lambda_n - \lambda_1) + \lambda_1)(\gamma(\lambda_1 - \lambda_n) + \lambda_n)$. It is a simple exercise to show that the optimal point γ is $1/2$, making (2.27) equal to $4\lambda_1\lambda_n/(\lambda_1 + \lambda_n)^2$, as required. $\qquad\qquad\square$

Inspect the right hand side of (2.25). Clearly, it equals $4\kappa(Q)/(1 + \kappa(Q))^2$, where $\kappa(Q) = \lambda_1/\lambda_n$ is known as the *condition number* of the matrix Q. Its range of values is the unit interval. In particular, the condition number equals zero if and only if Q is singular, and it equals one if and only if $Q = aI$ for some $a > 0$. Thus, the condition number is used to define the level of singularity of a matrix. Note that $\kappa(Q)$ is scale invariant, namely $\kappa(aQ) = \kappa(Q)$ for any scalar $a > 0$. Specifically and by definition, the closer it is to zero, the more singular Q is.

By Kontorovitch inequality (2.25) we get

$$1 - \frac{||x||^4}{(x^T Q x)(x^T Q^{-1} x)} \leq 1 - \frac{4\lambda_1\lambda_n}{(\lambda_1 + \lambda_n)^2} = \frac{(\lambda_n - \lambda_1)^2}{(\lambda_1 + \lambda_n)^2} = \frac{(1 - \kappa(Q))^2}{(1 + \kappa(Q))^2}.$$

Denote $Q^{-1}b$ by x^* and note that x^* is the unique optimal point for the function $\frac{1}{2}x^T Q x - b^T x + c$. For any $x \in R^n$, quantify its goodness as an approximation to x^* by $E(x)$, where $E(x) = (x - x^*)^T Q(x - x^*)$. As Q is positive, we know that $E(x) \geq 0$, and that it is an equality if and only if $x = x^*$.

Theorem

Theorem 2.6. *Let* x_i, $i \geq 0$, *be the sequence generated by applying the steepest descent method to the quadratic function* $\frac{1}{2}x^T Q x - b^T x + c$. *Then,*

$$E(x_{i+1}) = \left(1 - \frac{||g(x_i)||^4}{(g(x_i)^T Q g(x_i))(g(x_i)^T Q^{-1} g(x_i))}\right) E(x_i)$$

$$\leq \frac{(\lambda_n - \lambda_1)^2}{(\lambda_1 + \lambda_n)^2} E(x_i) = \frac{(1 - \kappa(Q))^2}{(1 + \kappa(Q))^2} E(x_i). \tag{2.28}$$

Proof. As $x_{i+1} = x_i - t^* g(x_i)$, simple algebra implies that

$$E(x_i) - E(x_{i+1}) = 2t^* g(x_i)^T Q(x_i - x^*) - 2t^* g(x_i)^T Q g(x_i).$$

Recall the expression for t^* (see (2.24)) and the fact that $Q(x_i - x^*) = g(x_i)$. Therefore,

$$E(x_i) - E(x_{i+1}) = \frac{2\|g(x_i)\|^4}{g(x_i)^T Q g x_i} - \frac{2\|g(x_i)\|^4}{g(x_i)^T Q g(x_i)} = \frac{\|g(x_i)\|^4}{g(x_i)^T Q g(x_i)}.$$

On the other hand,

$$x_i - x^* = Q^{-1} Q x_i - Q^{-1} b = Q^{-1}(Q x_i - b) = Q^{-1} g(x_i).$$

Hence,

$$E(x_i) = (x_i - x^*)^T Q(x_i)(x_i - x^*) = g(x_i)^T Q^{-1} g(x_i).$$

This completes the proof for the equality part of the theorem. The final inequality is now immediate from Lemma 2.1. $\square$

Going back to the motivation behind Theorem 2.6, we can say that when applying the steepest descent algorithm, the quality of the approximation at each iteration, measured by $(x - x^*)^T \nabla^2 f(x^*)(x - x^*)$, improves by a factor of approximately

$$\frac{(1 - \kappa(\nabla^2 f(x^*)))^2}{(1 + \kappa(\nabla^2 f(x^*)))^2}.$$

The closer one is to the optimal solution, the better this approximation is.

Exercise 2.6

Consider the minimization of the quadratic function $f(x) = \frac{1}{2} x^T Q x - b^T x$.

(a) What is $\nabla f(x)$? What is the steepest descent direction in x?

(b) Denote by d the steepest descent direction in x. Show that

$$\frac{d^T d}{d^T Q d} = \arg\min_{\alpha \in R} f(x + \alpha d).$$

(c) Let d_k and d_{k+1} be two consecutive directions generated by the steepest descent method for solving f(x). Show that d_k and d_{k+1} are orthogonal.

Exercise 2.7

Let $f(x) : R \rightarrow R$. We are looking for the polynomial $p(x)$ of degree n, which solves the following optimization problem:

$$\min_{p(x)} \int_0^1 (f(x) - p(x))^2 dx.$$

Show that it is an unconstrained quadratic programming problem with $n+1$ decision variables. In particular, find the Q matrix, the b vector, and the free coefficient.

Exercise 2.8

Consider the following unconstrained quadratic program:

$$\min_{x \in R^n} \frac{1}{2} x^T Q x - b^T x$$

for some square, symmetric, and positive matrix Q. Let $v_1, \ldots, v_n$ be a set of orthogonal eigenvectors Q. Suppose that for some x_0, the gradient at that point belongs to the subspace spanned by $v_1, \ldots, v_m$: $Sp\{v_1, \ldots, v_m\}$ for some $m < n$. Suppose that an iteration of the steepest descent algorithm was executed on this x_0. Denote by x' the resulting vector.

(a) Show that the gradient of the objective function at the new point x' also belongs to the subspace $Sp\{v_1, \ldots, v_m\}$.
(b) State your opinion: Is item (a) a strong or a weak property of the steepest descent algorithm?

2.5 The Stochastic Steepest Descent Method

Consider the example given in Section 2.3.5. We can see that the two optimization problems given in (2.20) and (2.21) are cases in which the function we optimize is in fact a sum of a number, m in these cases, of functions. When it comes to applying the steepest descent method, this is not an issue, at least not in theory: Just treat the sum of the functions as a single function. Yet, in cases where a large number of functions compose the sum, running the algorithm as it is can be computationally demanding per iteration. A practical idea here is that at any given iteration, only a sample of the m functions is used, and then the function we optimize in this iteration is the summation in the sample. In the context of neural networks, we can say that only a sample of the data is used in each iteration during the training procedure.

There are various strategies for selecting the sample. In the case where m is a mid-size number of functions and the sample size is k, then one can arrange the $m!/(k!(m-k)!)$ possible samples in some order and go through all of them in that order a number of times. However, when m is large, it is not ruled out that the number of iterations is larger than the possible samples. Here, the strategy is to collect random samples, each of size k, and hence the terminology of *stochastic steepest descent* method or *stochastic gradient descent* method. Note that there is nothing wrong with the choice of $k = 1$. Usually, we give all possible samples equal probabilities, and consecutive samples are drawn independently. Note that the sampled gradient times m/k is a multidimensional random variable whose expected value coincides with the gradient of the original function. Obviously, the improvement gained at each iteration is usually smaller than the one achieved by the original method as indeed the progress is not along the steepest descent direction. However, for a given computational budget, one can do more iterations, and hence one can be better off at the end of the day.

2.6   Minimum, Maximum, or Neither: Second-Order Conditions

Suppose that x^* does not meet the FOCs for optimality for the function $f(x) : R^n \to R$. Then, it is not a local minimum or maximum point. Thus, in our search for optimal points, we limit ourselves to points x^* that are *stationary*, namely $\nabla f(x^*) = \underline{0}$. There are two clear such cases: If $\nabla^2 f(x^*)$ is positive (respectively, negative), then it is a local minimum (respectively, maximum) point. Otherwise, we cannot say anything definite. This is stated formally as follows.

Theorem

Theorem 2.7. *Let X be a set in which x^* is an internal point. Also, let $f(x) : X \to R$. Assume that x^* meets the FOCs for optimality. Then a sufficient condition for a local minimum (respectively, maximum) point at x^* is that $\nabla^2 f(x^*)$ is positive (respectively, negative).*

Proof. Let $d \in R^n$ be some feasible direction at point x, namely for ϵ small enough $x^* + \epsilon d \in X$. Then, for any $\epsilon > 0$,

$$f(x^* + \epsilon d) = f(x^*) + \epsilon \nabla f(x^*)^T d + \frac{1}{2}\epsilon^2 d^T \nabla^2 f(y)d,$$

for some $y = x^* + \alpha \epsilon d$ with $0 \leq \alpha \leq 1$. Since the second term above equals zero and for ϵ small enough the third term is positive (respectively, negative) in the case where $\nabla^2 f(x^*)$ is positive (respectively, negative), our proof is complete.[k] □

Definition

Definition 2.6. For a function f, a point x is said to meet the second-order conditions (SOCs) if the matrix $\nabla^2 f(x)$ is positive.

Note that the SOCs basically say that at some (maybe small) neighborhood of x^*, the function is strictly convex (respectively, concave). Thus, the SOCs are automatically met in the case where $f(x)$ is strictly convex (for the local minimum point) or strictly concave (for the local maximum point). To summarize, in the case of a strictly convex or concave function, the FOCs are sufficient for local optimization. But in this case we can say much more: The optimal point is a global one. This can be stated formally as follows

---

Theorem

Theorem 2.8. *Let $f : X \to R$ be a convex (respectively, concave) function over a convex set X. Then, if x^* meets the FOCs for optimality, then x^* is a global minimum (respectively, maximum) point for $f(x)$ over X.*

Proof. Assume that $f(x)$ is convex (respectively, concave). Let $y \in X$ and assume that $y \neq x^*$. Then, for $z = \alpha x^* + (1 - \alpha)y$ for some α, $0 \leq \alpha \leq 1$,

$$f(y) = f(x^*) + \nabla f(x^*)^T (y - x^*) + \frac{1}{2}\epsilon^2 (y - x^*)^T \nabla^2 f(z)(y - x^*).$$

---

[k]An upper bound on ϵ for the validity of this statement is a function of both x^* and d.

The second term above equals zero while the third is non-negative (respectively, non-positive).   □

Are the SOCs for optimality stated in Theorem 2.7 also necessary? The answer is "almost". Of course, the FOCs are still necessary. Yet, in order to meet the SOCs, $\nabla^2 f(x^*)$ needs to be non-negative (respectively, non-positive). Put differently, in the existence of a direction $d \neq \underline{0}$ with $d^T \nabla^2 f(x^*)d < 0$ (respectively, $d^T \nabla^2 f(x^*)d > 0$) refutes a minimization (respectively, maximization) claim with respect to x^*. We leave the proof of this claim to the reader. Moreover, if $\nabla^2 f(x^*)$ is positive (respectively, negative) and not only non-negative (respectively, non-positive), the minimum (respectively, maximum), in case of a convex function, is global.

2.6.1 *Examples*

Example 1 (The least squares solution (cont.)). This problem was defined in Section 2.3.2. The Hessian matrix in this case is $A^T A$, regardless of the value of x. This is clearly a symmetric (namely square) matrix. Moreover, since $x^T A^T A x = ||Ax||^2$ for any $x \in R^n$, this Hessian is non-negative. Moreover, by definition in the case where A is a full-rank matrix, $Ax = \underline{0}$ if and only if $x = \underline{0}$, implying that $A^T A$ is positive and that the function optimized here is strictly convex. Thus, the solution x^* derived above is the unique global minimum, as claimed there.

Example 2 (The MLE for normal random variables (cont.)). This problem was defined in Section 2.3.1. Taking derivatives in (2.10) and (2.11) with respect to μ and σ^2, we get that the Hessian of the log likelihood function equals

$$-\frac{1}{\sigma^2} \begin{pmatrix} 1 & \frac{1}{\sigma^2}\sum_{i=1}^{n}(X_i - \mu) \\ \frac{1}{\sigma^2}\sum_{i=1}^{n}(X_i - \mu) & \frac{1}{\sigma^4}\sum_{i=1}^{n}(X_i - \mu)^2 - \frac{n}{2\sigma^2} \end{pmatrix}. \qquad (2.29)$$

Substituting here $\hat{\mu}$ and $\hat{\sigma}^2$ for μ and σ^2, respectively, we get that the Hessian matrix in the derived stationary point equals

$$-\frac{1}{\hat{\sigma}^2} \begin{pmatrix} 1 & 0 \\ 0 & \frac{n}{2\hat{\sigma}^2} \end{pmatrix}.$$

This is a negative matrix, implying that the stationary point $(\hat{\mu}, \hat{\sigma}^2)$ is a local maximum point. In fact, it is possible to show that the Hessian matrix (2.29) is negative for any pair of μ and $\sigma^2 > 0$, implying that the function is convex. Hence, the derived point is a global maximum, making the claimed point an MLE, as required.

2.7 Newton's Method

The steepest descent method is based on the idea of looking for the most promising direction to pursue a better solution, given the current approximation. Next, we deal with Newton's method. As always, the idea is to look locally, but now one approximates the function to be optimized using a quadratic function that gives us a good local approximation to $f(x)$ around a current approximation to the optimal point, and then looks for the minimum point of this quadratic function instead. We next give details about this procedure. In particular, we show that it implicitly suggests a direction to move along, and a line search can be applied to improve upon the plain vanilla procedure. Yet, the suggested direction is guaranteed to be a descent only if some mild conditions are met by the function $f(x)$.

Let $f(x) : R^n \to R$. Also, let x_k be some approximation to x^*. Define the following quadratic function:

$$q(x) = f(x_k) + \nabla f(x_k)^T (x - x_k) + \frac{1}{2}(x - x_k)^T \nabla^2 f(x_k)(x - x_k).$$

It is possible to see that

- $q(x_k) = f(x_k)$,
- $\nabla q(x_k) = \nabla f(x_k)$,
- $\nabla^2 q(x_k) = \nabla^2 f(x_k)$.

In other words, $q(x)$ is a quadratic function that, coupled with its first two derivatives, coincides with $f(x)$ at $x = x_k$. In other words, $g(x)$ is the quadratic fitting for $f(x)$ around x_k. Similarly, it can be seen as the result of ignoring the third and higher derivatives in the Taylor series approximation to $f(x)$ around x_k.

The function $q(x)$ can be expressed as $q(x) = \frac{1}{2}x^T Q x - b^T x + c$, where $Q = \nabla^2 f(x_k)$ and $b = -\nabla f(x_k) + \nabla^2 f(x_k)x_k$, where the value for c is immaterial. Thus, assuming that Q^{-1} exists, the stationary point of $q(x)$ is $Q^{-1}b$, namely

$$[\nabla^2 f(x_k)]^{-1}(-\nabla f(x_k) + \nabla^2 f(x_k)x_k) = x_k - [\nabla^2 f(x_k)]^{-1}\nabla f(x_k).$$

Newton's method is now as follows:

Input: $x_0 \in R^n$, $\epsilon > 0$.

Output: An approximation to x^*.

Initialization: $k = 0$.

Iterations: Let $x_{k+1} = x_k - [\nabla^2 f(x_k)]^{-1} \nabla f(x_k)$.

If $||x_{k+1} - x_k|| \leq \epsilon$ STOP

$k \leftarrow k + 1$. REPEAT

There are a few issues here. First, is x_{k+1} indeed the minimum of $q(x)$? The answer is positive if and only if $\nabla^2 f(x_k)$ is non-negative. In the case where $\nabla^2 f(x_k)$ is negative, we get in fact, a maximum point for $q(x)$. Of course, it is possible that $\nabla^2 f(x_k)$ is neither positive nor negative. Second, are we at least guaranteed that $f(x_{k+1}) < f(x_k)$? The answer is "not necessarily". After all, $q(x)$ is not $f(x)$, and who knows how the function $f(x)$ truly behaves. This is the case even if $\nabla^2 f(x_k)$ is positive. In order to circumvent the latter issue stated above, we add the following feature in the case where $\nabla^2 f(x_k)$ is positive. Specifically, note that $-[\nabla^2 f(x_k)]^{-1} \nabla f(x_k)$ defines a descent direction of the function $f(x)$ at the point x_k. This is the case since[1] $-\nabla f(x_k)^T [\nabla^2 f(x_k)]^{-1} \nabla f(x_k) < 0$. Thus, Newton's method can be interpreted as taking a step of size 1 along a specific descent direction, $d_k = -[\nabla^2 f(x_k)]^{-1} \nabla f(x_k)$, usually referred to as *Newton's direction*.

If needed, this direction can be normalized to be with a norm of one. One can apply a one-dimensional line search along this direction, looking for the optimal step size. Thus, define t^* as $\arg\min_{t \geq 0} f(x_k + td_k)$ and redefine x_{k+1} accordingly. The updated version of the algorithm is then:

Input: $x_0 \in R^n$, $\epsilon > 0$.

Output: An approximation to x^*.

Initialization: $k = 0$.

Iterations: Let

$$d_k = -[\nabla^2 f(x_k)]^{-1} \nabla f(x_k). \tag{2.30}$$

Let $t^* = \arg\min_{t \geq 0} f(x_k + td_k)$.

Let $x_{k+1} = x_k + t^* d_k$.

If $||x_{k+1} - x_k|| \leq \epsilon$ STOP

$k \leftarrow k + 1$. REPEAT

---

[1]Recall that a square matrix is positive if and only if its inverse is positive.

Third, does Newton's method guarantee local convergence to a point x^* with $\nabla f(x^*) = \underline{0}$? The answer is positive as long as $\nabla^2 f(x^*)$ is positive. Moreover, in this case, under mild conditions on the third derivative of $f(x)$ at the point x^*, the rate of convergence equals 2. This rate of convergence needs to be compared with the rate of convergence of 1 that is achieved by the steepest descent method. The proof of this rate of convergence, which is guaranteed by Newton's method, is similar to the proof for the case where $n = 1$ given in Section 1.7 above and is omitted for brevity. The interested reader can find a proof in [5, pp. 96–98].

The fourth issue is computational complexity. This is an important practical issue, and in our case, it has to do with inverting a matrix at each step. There is a vast literature on complexity, and in particular, on the complexity required for inverting matrices. Yet, for our purposes, it is enough to assume that inverting a square matrix requires an order of magnitude of n^3, while multiplying a vector by a matrix requires an order of magnitude n^2. Admittedly, more advanced procedures for inverting less complex matrices exist. Note that instead of inverting the matrix $\nabla^2 f(x^k)$, which is required for the execution of Newton's method, it is possible to solve for d_k the set of linear equations

$$\nabla^2 f(x^k)d_k = -\nabla f(x^k).$$

This can be done with any linear equation solver. For example, in the case where $\nabla^2 f(x^k)$ is known to be a positive matrix and the the function $f(x)$ is convex, one can use the Cholesky factorization technique. For more on this technique see, e.g., [14, p. 150].

In the next chapter, we introduce techniques that make it possible to replace $[\nabla^2 f(x_k)]^{-1}$ above by some other matrix, possibly one that approximates the term well. Moreover, computing it requires an effort of an order of magnitude of n^2. In such a case, some of the optimality features of Newton's method will be lost, but as we start converging to x^*, the corresponding $[\nabla^2 f(x_k)]^{-1}$ will not vary much, and therefore a computationally simple update of the (approximate) inverse will be almost as good. In fact, since we will perform fewer computations at each iteration, we will be able to perform more iterations at the same cost and perhaps even be better off. We give some details on such techniques in the next section.

Remark. In comparing the steepest descent direction with Newton's direction, we can say that in order to get Newton's direction, we need to rotate the steepest descent direction by $[\nabla^2 f(x_k)]^{-1}$. Of course, in either case, we assume that scaling is used, and the norms of the directions are with a value of 1.

Exercise 2.9

Let $g_i(x)$, $1 \leq i \leq n$, be n functions from R^n to itself. Our goal is to find a solution $x \in R^n$ that satisfies $g_i(x) = 0$, $1 \leq i \leq n$.

(a) Show that this problem is equivalent to solving the unconstrained optimization problem

$$\min_{x \in R^n} \sum_{i=1}^{n} g_i^2(x).$$

(b) Explain how the steepest descent method is applied to the unconstrained optimization problem.

(c) Explain how Newton's method is applied to the unconstrained optimization problem.

(d) What is your opinion on the idea that all of the above will be replaced such that the objective function to be minimized is

$$\min_{x \in R^n} \sum_{i=1}^{n} |g_i(x)|?$$

Exercise 2.10

Let $f_i(x) : R^n \to R, 1 \leq i \leq m$, be a set of real m functions. For some given real numbers c_i, $1 \leq i \leq m$, define the following function

$$g(x) = \sum_{i=1}^{m} (f_i(x) - c_i)^2,$$

and consider the following optimization problem:

$$\min_{x \in R^n} g(x).$$

(a) What are the gradient and Newton's direction of the function $g(x)$ at some given point x?

(b) A suggested iterative method for solving for the value minimizing $g(x)$ is

$$x_{k+1} = \min_{x \in R^n} \sum_{i=1}^m \left(f_i(x_k) + \nabla f_i(x_k)^T (x - x_k) - c_i \right)^2.$$

Is this the steepest descent method in disguise? What is the idea behind that method?

(c) Show that x_{k+1} is a solution to the quadratic optimization of the type $arg \max_x x^T Q x - 2b^T x$. What are Q and b?

2.8 Quasi-Newton Methods

Iterative procedures are usually based on three factors: (i) current approximation x, (ii) a descent direction d, and (iii) a step size t. In particular, the next approximation is

$$x \to x + td.$$

Moreover, using the steepest descent direction $-\nabla f(x)$ as a point of departure, one can look at d as $d = -S\nabla f(x)$ for some choice of a matrix $S \in R^{n \times n}$. Indeed, by the steepest descent algorithm, $S = I$, while by Newton's method, $S = [\nabla^2 f(x)]^{-1}$. These two choices can be looked at as two extreme choices. Specifically, the former is the simplest choice, while the latter is more computationally demanding. The former choice has a linear convergence rate, while the latter has a (local) quadratic rate. Do we have other "intermediate" options? Enter quasi-Newton methods. The seeming trade-off in the selection of S is between computational simplicity and closeness to $[\nabla^2 f(x)]^{-1}$. Usually, the computational complexity of the suggested matrix S is $O(n^2)$ (as opposed to the computational complexity of $[\nabla^2 f(x)]^{-1}$, which, under standard procedures, is $O(n^3)$). The sequence of generated matrices should converge to $\nabla^2 f(x^*)$, which in itself does not guarantee a quadratic rate of convergence.

In the next two subsections, we give details on possible choices for the matrix S and explain them. They all share the following idea: Define by x_i, $i \geq 0$, the to-be-generated sequence of approximations to x^*, and assume that $\lim_{i \to \infty} x_i = x^*$. Suppose that $\nabla^2 f(x_i)^{-1}$ is known. At the next iteration, you would ideally compute $\nabla^2 f(x_{i+1})^{-1}$, but this is computationally demanding. Yet, as the algorithm progresses, the series of these inverses does not change. This raises the question of whether it is worth the effort to compute the inverses precisely or whether one should be satisfied with a good approximation whose computational effort is light. The designed algorithms are as Newton's, except now some matrix, say S_k, replaces $\nabla^2 f(x_k)$ in (2.30).

The following algorithm is a good example of this approach.

2.8.1 *Rank-one updates*

Input: $x_0 \in R^n$, $H_0 = I$, $g_0 = \nabla f(x_0)$, $\epsilon > 0$.
Output: An approximation of x^*.
Initialization: $k = 0$.
Iterations:
Let $d_k = -H_k g_k$.
Let $t^* = \arg\min_{t \geq 0} f(x_k + t d_k)$.
Let

$$x_{k+1} = x_k + t^* g_k,$$

$$g_{k+1} = \nabla f(x_{k+1}),$$

$$q_k = g_{k+1} - g_k,$$

and

$$p_k = t^* d_k.$$

Let

$$H_{k+1} = H_k + \frac{(p_k - H_k q_k)(p_k - H_k q_k)^T}{q_k^T (p_k - H_k q_k)}. \tag{2.31}$$

If $||x_{k+1} - x_k|| \leq \epsilon$ STOP
$k \leftarrow k + 1$ REPEAT

Note that in (2.31) we have the outer product of a vector, $p_k - H_k q_k$, divided by a constant (which is the inner product between two vectors). This results in a rank-one matrix, whose computational complexity is $O(n^2)$. This part replaces the computation of $\nabla^2 f(x_k)^{-1}$, which is computationally more demanding. Before explaining the rationale behind this update, note that if H_k is a symmetric matrix, then H_{k+1} is also a symmetric matrix. This choice of H_0 guarantees symmetry, but any other choice for a symmetric H_0 possesses all properties described above. On the other hand, the positivity of H_k is not sufficient to guarantee the positivity of H_{k+1}, and indeed, counterexamples exist. This issue is tackled in the next section.

The idea behind the rank-one update algorithm is as follows. First, using a second-degree approximation, we get that for any reasonable algorithm,

$$\nabla f(x_{i+1}) - \nabla f(x_i) \approx \nabla^2 f(x_i)(x_{i+1} - x_i).$$

As the series $\nabla^2 f(x_i)^{-1}$ approaches convergence, these matrices do not vary by much. Since $\nabla^2 f(x_i)^{-1}$ is positive, if we replace it with a constant matrix to be called here H_k, and assume that H_k is used for k consecutive steps for some $k \leq n - 1$, then,

$$x_{i+1} - x_i \approx H_k(\nabla f(x_{i+1}) - \nabla f(x_i)), \quad 1 \leq i \leq k.$$

Now, let us reverse the logic and assume that the above approximation is exact, namely

$$x_{i+1} - x_i = H_k(\nabla f(x_{i+1}) - \nabla f(x_i)), \quad 1 \leq i \leq k. \tag{2.32}$$

Put differently, for any $x_{i+1} - x_i$ and $\nabla f(x_{i+1}) - \nabla f(x_i)$, $1 \leq i \leq k \leq n-1$, look for a matrix H_k that meets (2.32). Unless some degeneracy exists, this is always possible since we have an undetermined set of k^2 equations with n^2 unknown variables. This idea is summarized in the following theorem.

> ## Theorem
>
> **Theorem 2.9.** *Denote $x_{i+1} - x_i$ by p_i and $\nabla f(x_{i+1}) - \nabla f(x_i)$ by q_i, $1 \leq i \leq n$, which are generated by the rank-one update algorithm. Assume that $p_i = H_k q_i$, $1 \leq i < k \leq n - 1$. Then, $p_i = H_{k+1} q_i$, $1 \leq i < k + 1 \leq n$. Moreover, the update given in (2.31) is the unique rank-one update having this property.*

Proof. Our proof will be constructive. Note that we are looking for a scalar α_k and a vector $z_k \in R^n$ such that

$$H_{k+1} = H_k + \alpha_k z_k z_k^T$$

satisfies

$$H_{k+1} q_i = p_i, \quad 1 \leq i \leq k. \tag{2.33}$$

Observe that the above set of equalities has one additional equality requirement on top of the $k - 1$ equalities assumed to be met by H_k. True, there is no uniqueness here because z_k can be multiplied by some constant c and

α_k divided by c^2 to compensate. Yet, one cannot assume without loss of generality that $\alpha_k = 1$ as its value can be negative and hence not be fully "absorbed" by z_k.

We first examine (2.33) for the case where $i = k$:

$$p_k = H_{k+1}q_k = (H_k + \alpha_k z_k z_k^T)q_k = H_k q_k + \alpha_k z_k z_k^T q_k,$$

which implies that

$$p_k - H_k q_k = \alpha_k z_k z_k^T q_k. \qquad (2.34)$$

Left multiplying this by q_k, we get that

$$q_k^T(p_k - H_k q_k) = \alpha_k (q_k^T z_k)^2,$$

and

$$\alpha_k = \frac{q_k^T(p_k - H_k q_k)}{(q_k^T z_k)^2}. \qquad (2.35)$$

Now right multiplying (2.34) by $(p_k - H_k q_k)^T$ (an outer product), we observe that

$$(p_k - H_k q_k)(p_k - H_k q_k)^T = \alpha_k^2 z_k z_k^T q_k q_k^T z_k z_k^T = \alpha_k^2 (z_k^T q_k)^2 z_k z_k^T,$$

and

$$\alpha_k z_k z_k^T = \frac{(p_k - H_k q_k)(p_k - H_k q_k)^T}{\alpha_k(z_k^T q_k)^2} = \frac{(p_k - H_k q_k)(p_k - H_k q_k)^T}{q_k^T(p_k - H_k q_k)},$$

where the last equality follows from (2.35). This completes the uniqueness part. The only thing left to show is that $H_{k+1}q_i = p_i$, $1 \le i < k$. Indeed,

$$H_{k+1}q_i = H_k q_i + \frac{p_k - H_k q_k}{q_k^T(p_k - H_k q_k)}(p_k^T q_i - q_k^T H_k q_i).$$

Note that in the last equality, we used the fact that H_k is symmetric. Finally, inspect the term $p_k^T q_i - q_k^T H_k q_i$. The assumption made in the theorem says that $H_k q_i = p_i$, and hence this difference boils down to $p_k^T q_i - q_k^T p_i$. This term equals zero since $p_k^T q_i = p_k^T H_k^{-1} p_i = q_k^T p_i$. This completes the proof. $\qquad \square$

We have thus explained the intuition behind the algorithm and stated its computational advantage. Yet, convergence, and to an optimal point, is not guaranteed. The reason is that the algorithm is based on directions which are not necessarily descent directions. In particular, it is possible to construct an example where $f(x_{i+1}) - f(x_i) > 0$. This limitation is removed in the next section.

2.8.2   Rank-two updates: the Davidon–Fletcher–Powell method

The rank-two update algorithm replaces (2.31) with

$$H_{k+1} = H_k + \frac{p_k p_k^T}{p_k^T q_k} - \frac{H_k q_k q_k^T H_k}{q_k^T H_k q_k}.$$

Note that now two rank-one matrices are added to H_k, leading to the name of this algorithm. This update comes with all the properties of the rank-one update. The added feature is that if one initializes with H_0, which is a positive matrix, the generated sequence of matrix H_k, $k \geq 1$, preserves this property. This fact in turn leads to d_k, $k \geq 1$, being descent directions. In particular, the sequence of generated objective function values is decreasing. We do not prove this claim. The reader is referred to [18, pp. 304–306], for a comprehensive proof. Note that the added computational burden is $O(n^2)$, which preserves the computational complexity of the rank-one update algorithm (but nevertheless doubles the computational burden at each iteration).

2.9   The Expectation-Maximization (EM) Algorithm

2.9.1   The algorithm

Revisiting Section 1.4, we let $L(y; \theta)$ be a likelihood function. Usually, $y \in R^n$ is a vector representing a sample realization, and θ is a parameter characterizing a family of distributions (which can be single- or multi-valued). Thus, $L(y; \theta)$ is the likelihood function or the density function (or a mixture of the two) of y when θ is the value for the parameter that governs the likelihood. We are interested in the MLE:

$$\hat{\theta} = \arg\max_{\theta} L(y; \theta), \tag{2.36}$$

which is a function of y. In many cases, this is a relatively easy problem. In particular, one differentiates (if possible) the likelihood function with respect to θ and finds a value of θ where the derivative equals zero. Two examples are given in Section 1.4. More examples can be found in any statistical theory textbook. One possible source is Section 7.2.2 of [19].

This section deals with such cases where computing the MLE is a hard task. However, assume for now that there exists another variable x, such that with y and given θ, has a joint likelihood function. Denote by $L(y, x; \theta)$ the resulting likelihood function, known in this context as the *complete likelihood* function. Assume now that finding the MLE, given (y, x), is easy; that is, finding

$$\arg \max_{\theta} L(y, x; \theta),$$

which is a function of both y and x, is a doable task, which, for example, can be done by a straightforward differentiation. The snag here is that this is not the maximization problem we are facing since the realized x is not given. The only thing that we know is the conditional distribution of x given y (which in turn is a function of θ), where y is the *data* and x is the *missing* data. The variables that correspond to the missing data are called *latent* variables.

The solution here is to use the expectation-maximization (EM) algorithm. It is not an off-the-shelf algorithm, due to the fact that it admits freedom in selecting the latent random variables, denoted by X. This is sometimes more a question of art, experience, and common sense than a question of science.

The algorithm itself runs as follows. Start with some θ_0, and generate the following sequence θ_j, $j \geq 1$, until some convergence is achieved:

$$\theta_{j+1} = \arg \max_{\theta} \int_x \log L(y, x; \theta) f_{X|Y=y, \theta_j}(x) \, dx, \quad j \geq 0,$$

or, put differently,

$$\theta_{j+1} = \arg \max_{\theta} \mathrm{E}_{X|Y=y, \theta_j}(\log L(y, X; \theta)). \tag{2.37}$$

In words, the procedure can be described in the following steps. First, let y be the given data and assume that they do not change during the run of the procedure. Second, write the log of the complete likelihood function for an arbitrary value for the parameter θ. Third, and this is the expectation part, find the expected value of this function (again, for any value of the parameter θ) with respect to the latent variables given the data and the current value of the parameter θ_j. Note that you still get a function of the parameter θ. Finally, and this is the maximization part, derive the value of the parameter θ that maximizes this function. Thus, θ_{j+1}.

You may wonder why the log is being introduced here and not another monotonic increasing function. In fact, why not use the likelihood function itself? The reason, which we state without a proof, is that $L(y; \theta_{j+1}) \geq L(y; \theta_j)$. In fact, any other strictly concave and monotonic increasing function, such as the log function, has this property.[m] These concavity and monotonicity assumptions imply that the generated sequence of likelihood values, at least in the limit, converges to a local maximum value of the original likelihood function. This convergence does not, in principle, guarantee the convergence of the series θ_j when j goes to infinity, but in practice, it is usually the case. In many cases, this local maximum is also global, and the convergence is to the MLE. Note that hence, if the input parameter is the MLE, then it is also the output parameter. In other words, the MLE is a fixed point of the function (in θ_0), $\arg\max_\theta \int_x \log L(y, x; \theta) f_{X|Y=y,\theta_0}(x)$.

Remark. A possible shortcut is to replace the right-hand side of (2.37) with

$$\arg\max_\theta \log L(y, \mathrm{E}(X|_{Y=y,\theta_j}; \theta)).$$

It is tantamount to replacing the expected value of a function with the function of the expected value. This leads to a suboptimal solution, and as opposed to the original version, convergence of the likelihood function is not guaranteed now. Yet, a possible advantage is that less effort is required per iteration.

2.9.2   *Example 1: censored exponential data*

Let $X_i \sim \exp(\lambda)$, $1 \leq i \leq n$, be n independent random variables. For some $T > 0$ let $Y_i = \min\{X_i, T\}$, $1 \leq i \leq n$; this is known as the *censored data.* What is the MLE for λ, given the sample Y_i, $1 \leq i \leq n$, assuming that T is known but not the X_i, $1 \leq i \leq n$? First,

$$L(Y_1, \ldots, Y_n; \lambda) = \prod_{i:Y_i<T} \lambda e^{-\lambda Y_i} \prod_{i:Y_i=T} e^{-\lambda T}.$$

We take the log of the above equality, differentiate it with respect to λ, set the resulting derivative to zero, solve for λ, and conclude that the MLE equals

$$\frac{r}{\sum_{i:Y_i<T} Y_i + (n-r)T}, \tag{2.38}$$

---

[m]The interested reader can find a proof in [11, p. 361].

where r is the number of observations out of Y_i, $1 \leq i \leq n$, which are smaller than T. This is not a typical example, as in this case, it is possible to easily derive the MLE. Yet, the purpose of this example is to exemplify the EM algorithm. Suppose now that the complete data are (X_i, Y_i), $1 \leq i \leq n$. In other words, X_i, $1 \leq i \leq n$, are the latent variables. The fact that in this specific example $X_i = Y_i$, where $Y_i \leq T$, is not an issue. The complete likelihood function is hence

$$L(Y_1, \ldots, Y_n, X_1, \ldots, X_n; \lambda) = \prod_{i:Y_i<T} \lambda e^{-\lambda Y_i} \prod_{i:Y_i=T} \lambda e^{-\lambda X_i}$$

$$= \lambda^n e^{-\lambda \sum_{i|Y_i<T} Y_i} e^{-\lambda \sum_{i|Y_i=T} X_i}.$$

Hence,

$$\log L(Y_1, \ldots, Y_n, X_1, \ldots, X_n; \lambda) = n \log \lambda - \lambda \sum_{i|Y_i<T} Y_i - \lambda \sum_{i|Y_i>T} X_i.$$
$$(2.39)$$

Note that the conditional distribution of X_i given $(Y_1, \ldots, Y_n)$, $1 \leq i \leq n$, equals the conditional distribution of X_i given Y_i, $1 \leq i \leq n$, which in turns equals

$$X_i = \begin{cases} Y_i, & \text{if } Y_i \leq T, \\ T + W_i, & \text{if } Y_i = T, \end{cases}$$

where $W_i \sim exp(\lambda)$, $1 \leq i \leq n$, are n independent random variables that are also independent of all of the other random variables defined so far. Note that the memoryless property of the exponential distribution is invoked here.

We are now ready to state the expectation part. The expected value of (2.39) given Y_i, $1 \leq i \leq n$, when the parameter governing the latent variables is λ_0, equals

$$n \log \lambda - \lambda \sum_{i|Y_i<T} Y_i - \lambda \sum_{i|Y_i>T} \left(T + \frac{1}{\lambda_0}\right)$$

$$= n \log \lambda - \lambda \sum_{i|Y_i<T} Y_i - \lambda(n-r)T - \frac{\lambda(n-r)}{\lambda_j}, \quad j \geq 0.$$

And now we turn to the maximization part. Take the derivative with respect to λ and denote by λ_1 the value for λ for which this derivative equals zero. We get

$$\frac{n}{\lambda_1} - \sum_{i|Y_i<T} Y_i - (n-r)T - \frac{n-r}{\lambda_0} = 0, \qquad (2.40)$$

from which we can easily derive λ_1 as a function of λ_0, and in fact λ_{j+1} as a function of λ_j. Specifically,

$$\lambda_{j+1} = \frac{n}{\sum_{i|Y_i<T} Y_i + (n-r)T + \frac{n-r}{\lambda_j}}, \qquad j \geq 0.$$

The theory tells us that regardless of λ_0, if the original to-be-maximized likelihood function possesses a unique maximum (which was proved above in (2.40) for this model) and if convergence is guaranteed (which was not argued for here), then

$$\lim_{j\to\infty} \lambda_j = \frac{r}{\sum_{i:Y_i<T} Y_i + (n-r)T},$$

which is the MLE (see (2.38)). It is also easy to check that $\lambda_1 = \lambda_0$ if and only if λ_0 equals the MLE. In other words, the MLE is the unique fixed point of the EM algorithm.

2.9.3 *Example 2: mixture of distributions*

Suppose that for three parameters p, $0 \leq p \leq 1$, and $\lambda_1, \lambda_2 > 0$, the density function of a single-value random variable Y is

$$f_Y(y) = p\lambda_1 e^{-\lambda_1 y} + (1-p)\lambda_2 e^{-\lambda_2 y}, \quad y \geq 0.$$

This is a mixture between two exponential densities. The generalization for mixing a larger number of exponential densities is immediate, and we keep it to two for ease of exposition. Of course,

$$L(Y_1,\ldots,Y_n; p,\lambda_1,\lambda_2) = \prod_{i=1}^{n} (p\lambda_1 e^{-\lambda_1 Y_i} + (1-p)\lambda_2 e^{-\lambda_2 Y_i}).$$

Finding the MLE for $(p, \lambda_1, \lambda_2)$ does not seem to be an easy task.

Let us complete the data so that the n independent random variables I_i, $1 \leq i \leq n$, are with $I_i = 1$ (if the first density "wins" the lottery regarding where Y_i is from) or $I_i = 0$ (otherwise). These are the latent variables that are independent and identically distributed à la Bernoulli with parameter p.

What we are in fact saying is that the conditional marginal distribution of I_i is as follows:

$$P(I_i = 1|(Y_1, \ldots, Y_n)) = \frac{p\lambda_1 e^{-\lambda_1 Y_i}}{p\lambda_1 e^{-\lambda_1 Y_i} + (1-p)\lambda_2 e^{-\lambda_2 Y_i}}, \quad 1 \leq i \leq n,$$

(2.41)

with an identical expression for the conditional expected value for I_i, $1 \leq i \leq n$. We also assume that all these indicator variables are conditionally independent. Note that the conditional distribution is defined only by Y_i (although in general it is a function of $(Y_1, \ldots, Y_n)$). Note also that when we think of a mixture, we think of sampling for the type first and then for the value, but there is nothing wrong if we sample first for the value and then for the type. For example, given the value, the lottery regarding the type needs to be done with the probabilities stated in (2.41).

The complete likelihood function is then

$$L(Y_1, \ldots, Y_n, I_1, \ldots, I_n; p, \lambda_1, \lambda_2) = \prod_{i|I_i=1} p\lambda_1 e^{-\lambda_1 Y_i} \prod_{i|I_i=0} (1-p)\lambda_2 e^{-\lambda_2 Y_i},$$

(2.42)

for which it is easy to derive the MLE. Specifically, its log is easily seen to equal

$$K \log p + (n-K) \log(1-p) + K \log \lambda_1 - \lambda_1 \sum_{i:I_i=1} Y_i + (n-K) \log \lambda_2 - \lambda_2 \sum_{i:I_i=0} Y_i,$$

(2.43)

where K is the number of I_i's with $I_i = 1$. This likelihood function is clearly maximized with

$$p = \frac{K}{n}, \quad \lambda_1 = \frac{K}{\sum_{i:I_i=1} Y_i}, \quad \text{and} \quad \lambda_2 = \frac{n-K}{\sum_{i:I_i=0} Y_i}.$$

(2.44)

Unfortunately, this is not the maximization problem we need to solve. Yet, it is an indication that the use of the EM algorithm with $(I_1, \ldots, I_n)$ as the choice for the set of latent variables is promising.

Recall that the log of the complete likelihood function given in (2.42) is in fact

$$K \log p + (n-K) \log(1-p)$$
$$+ K \log \lambda_1 - \lambda_1 \sum_{i=1}^{n} I_i Y_i + (n-K) \log \lambda_2 - \lambda_2 \sum_{i=1}^{n} (1-I_i) Y_i. \quad (2.45)$$

Therefore, we apply the EM algorithm to it. Let p^j, λ_1^j, and λ_2^j be the current approximations for the parameters we are after. Denote the conditional expected value for the latent variable I_i by $W_i(Y_i; p^j, \lambda_1^j, \lambda_2^j)$. These conditional probabilities are stated in (2.41). Note that in principle we should have $(Y_1, \ldots, Y_n)$ instead of Y_i here. Then,

$$\mathrm{E}(K | (\underline{Y}; p^j, \lambda_1^j, \lambda_2^j)) = \sum_{i=1}^{n} W_i(Y_i; p^j, \lambda_1^j, \lambda_2^j).$$

Thus, the expectation part of the EM algorithm is now exactly the same as (2.43), where K is replaced by its conditional expected value $\Sigma_{i=1}^{n} W_i(Y_i; p^j, \lambda_1^j, \lambda_2^j)$. Thus, the expectation part ends with

$$K^j \log p + (n - K^j) \log(1 - p) + K^j \log \lambda_1 - \lambda_1 \sum_{i=1}^{n} W_i(Y_i; p^j, \lambda_1^j, \lambda_2^j) Y_i$$

$$+ (n - K^j) \log \lambda_2 - \lambda_2 \sum_{i=1}^{n} (1 - W_i(Y_i; p^j, \lambda_1^j, \lambda_2^j)) Y_i,$$

where $K^j = \Sigma_{i=1}^{n} W_i(Y_i; p^j, \lambda_1^j, \lambda_2^j)$.

Now, we maximize this with respect to p, λ_1, λ_2 and get $p^{j+1}, \lambda_1^{j+1}, \lambda_2^{j+1}$. Thus, we mimic (2.43), where K is replaced with K^j which is its conditional on the data expected value under the current set of approximations. In summary, the next set of new approximations is

$$p^{j+1} = \frac{\sum_{i=1}^{n} W_i(Y_i; p^j, \lambda_1^j \lambda_2^j)}{n},$$

and

$$\lambda_1^{j+1} = \frac{\sum_{i=1}^{n} W_i(Y_i; p^j, \lambda_1^j, \lambda_2^j)}{\sum_{i=1}^{n} W_i(Y_i; p^j, \lambda_1^j, \lambda_2^j) Y_i}, \quad \lambda_2^{j+1} = \frac{n - \sum_{i=1}^{n} W_i(Y_i; p^j, \lambda_1^j, \lambda_2^j)}{\sum_{i=1}^{n} (1 - W_i(Y_i; p^j, \lambda_1^j, \lambda_2^j)) Y_i}.$$

Exercise 2.11

Let $X_i \sim Pois(\lambda_i)$ and $Y_i \sim Pois(\beta \lambda_i)$, $1 \leq i \leq n$, be $2n$ independent random variables that follow Poisson distributions, where $\lambda_i > 0, 1 \leq i \leq n$, and $\beta > 0$ are $n + 1$ parameters to be estimated.[n]

1. What is the (complete) likelihood function?

---

[n]This exercise appears in [11, p. 359].

2. Prove that the MLEs are

$$\hat{\beta} = \frac{\sum_{j=1}^{n} Y_i}{\sum_{j=1}^{n} X_i} \quad \text{and} \quad \hat{\lambda}_i = \frac{X_i + Y_i}{\hat{\beta} + 1}, \quad 1 \le i \le n.$$

3. Suppose that X_1 is a missing datum. Show that the MLEs for λ_i, $1 \le i \le n$, and β, based on the incomplete data, solve the following (nonlinear) equations:

$$\hat{\beta} = \frac{\sum_{i=1}^{n} Y_i}{\sum_{i=1}^{n} \hat{\lambda}_i}, \quad \hat{\lambda}_1 = \frac{Y_1}{\hat{\beta}}, \quad \text{and} \quad \hat{\lambda}_i = \frac{X_i + Y_i}{\hat{\beta} + 1}, \quad 2 \le i \le n. \quad (2.46)$$

4. Design the resulting EM algorithm (for the incomplete data). In particular, do the following:

   (a) For the expectation part, state the expected log likelihood function.
   (b) For the maximization part, show that if the current approximations for the MLEs are $\hat{\beta}^0$ and $\hat{\lambda}_i^0$, $1 \le i \le n$, then the next ones are

   $$\hat{\beta}^1 = \frac{\sum_{i=1}^{n} Y_i}{\hat{\lambda}_1^0 + \sum_{i=2}^{n} X_i}, \quad \hat{\lambda}_1^1 = \frac{\hat{\lambda}_1^0 + Y_1}{\hat{\beta}^1 + 1}, \quad \text{and}$$

   $$\hat{\lambda}_i^1 = \frac{X_i + Y_i}{\hat{\beta}^1 + 1}, \quad 2 \le i \le n. \quad (2.47)$$

   Note that $\hat{\lambda}_1^0$ replaces here the missing datum of X_1.

5. Show that the fixed point of the iterative procedure stated in (2.47) solves the equations given in (2.47).

Part II
Constrained Optimization

Chapter 3

Optimization Under Equality Constraints: Special Cases

3.1 Introduction

The simplest optimization problem with one equality constraint is

$$\min_{x \in R^n} \sum_{i=1}^{n} a_i x_i,$$

$$\text{s.t.} \ \sum_{i=1}^{n} b_i x_i = c,$$

$$x_i \geq 0, \quad 1 \leq i \leq n.$$

Assume that $a_i > 0$ and $b_i > 0$, $1 \leq i \leq n$, and $c > 0$. It can be looked as an allocation problem in which an amount c of some resource has to be allocated among n activities. One unit of activity i consumes b_i units of the resource and adds b_i to the common cost, $1 \leq i \leq n$. Look at the index $i^* = \arg\min_{i=1}^{n} a_i/b_i$, and for simplicity, assume that this index is uniquely defined. It is easy to see that one wishes to increase x_{i^*} as much as possible, at the expense of the other variables. Yet, these $n-1$ variables are bounded from below by zero. Thus, the optimal solution is

$$x_{i^*} = \frac{c}{b_{i^*}} \quad \text{and} \quad x_i = 0, \quad 1 \leq i \neq i^* \leq n.$$

3.2 Allocation Problems

Let $f_i(x) : R^1 \to R^1$, $1 \leq i \leq n$, be n single-variable functions. Assume that they are monotone increasing and strictly convex, namely $f_i'(x) > 0$

and $f_i''(x) > 0$, $1 \leq i \leq n$. Finally, assume that $\lim_{x \to \infty} f_i'(x) = \infty$ and $\lim_{x \to -\infty} f_i'(x) = -\infty$, $1 \leq i \leq n$.

Consider the following problem:

$$\min_{x_i, 1 \leq i \leq n} \sum_{i=1}^{n} f_i(x_i),$$

$$\text{s.t.} \quad \sum_{i=1}^{n} a_i x_i = c,$$

$$x_i \geq 0, \quad 1 \leq i \leq n,$$

for some $n + 1$ positive numbers, c, and a_i, $1 \leq i \leq n$. Assume without loss of generality that $c = 1$, $a_i = 1$, and $f_i(0) = 0$, $1 \leq i \leq n$. Note that, were this not the case, then the change of variables $y_i = a_i x_i / c$ and the replacement of $f_i(x_i)$ by $g_i(y_i) = f_i(cy_i/a_i) - f_i(0)$, $1 \leq i \leq n$, would lead to an equivalent problem that meets all the above assumptions. Thus, the problem becomes

$$\min_{x_i, 1 \leq i \leq n} \sum_{i=1}^{n} f_i(x_i), \qquad (3.1)$$

$$\text{s.t.} \quad \sum_{i=1}^{n} x_i = 1,$$

$$x_i \geq 0, \quad 1 \leq i \leq n,$$

where all functions $f_i(x)$, $1 \leq i \leq n$, share the same value of zero where $x_i = 0$, $1 \leq i \leq n$.

Note that, the objective function here is *separable* in the sense that the decision variables have minimal interaction. Namely, each variable is transformed using the functions $f_i(x_i)$, $1 \leq i \leq n$, and the transformed values are added up. This is a simple operation that should, or at least is desired to, result in a relatively easy way to solve the stated optimization problem. This, when coupled with a single linear constraint, is indeed the case as we outline below.

Before dealing with the problem stated in (3.1), note that, without the non-negativity constraints in it, the problem would have been rather simple. Specifically, a necessary condition for optimality is that all derivatives at the optimal point should be equal. To see this, suppose, for the contrary that $f_1'(x_1) < f_2'(x_2)$. Then, it would have been worthwhile to increase x_1 gradually and decrease x_2 accordingly, while still meeting the equality constraint, as this would lead to a reduction in the objective function. Note that, due to the convexity of both functions, the size of this change is bounded until x_i reaches the point where both derivatives coincide. The assumption that the derivatives go to infinity (minus infinity, respectively) when the variables go to infinity (minus infinity, respectively) implies that this process is doable. In summary, a necessary condition for x_i, $1 \leq i \leq n$, to be the optimal solution, is that $f(x_i) = c$, $1 \leq i \leq n$, for some constant c. Another rather trivial comment is that without the constraint $\Sigma_{i=1}^{n} x_i = 1$, the optimal solution would, of course, be $x_i = 0$, $1 \leq i \leq n$.

The takeaway lesson of this simplified version of the problem is that a necessary condition for optimality is that if $x_1, x_2 > 0$, then $f'_1(x_1) = f'_2(x_2)$. Moreover, if one considers a feasible solution with $f'_1(x_1) < f'_2(x_2)$, then one would wish to increase x_1 and accordingly, decrease x_2, until these two derivatives coincide. Yet, the constraint $x_2 \geq 0$ might be on one's way in decreasing x_2, and possibly one needs to stop when $x_2 = 0$. From this, we learn that a necessary condition for optimality when $x_1 > 0$ and $x_2 = 0$ is that $f'_1(x_1) \leq f'_2(0)$ (and not $f'_1(x_1) = f'_2(0)$). We next show how this insight leads to a procedure for computing the optimal solution.

3.2.1 *Gradual increasing of variables*

Look at the trial $x_i = 0$, $1 \leq i \leq n$. It is a great choice, but it does not meet the constraint $\Sigma_{i=1}^{n} x_i = 1$. Thus, at least one of the variables needs to be larger than zero. Assume without loss of generality that $f'_1(0) \leq f'_2(0) \leq \cdots \leq f'_n(0)$, and, for simplicity, all inequalities are strict. Thus, the first variable to be increased from zero is x_1: The increase in the objective function due to an (infinitesimal) increase in one of the variables is minimal here. How much should x_1 be increased? Note that the increase in x_1 entails an increase in $f'_1(x_1)$. Hence, the increase should be up to 1 or up to the point $x_1 < 1$, where $f'_1(x_1) = f'_2(0)$, whichever comes first. In the former case, we conclude that $x_1 = 1$ and $x_i = 0$, $2 \leq i \leq n$, is the optimal solution. In the latter case, we are not done yet. We still need to increase x_2 and x_1 while keeping the derivatives of these two functions equal. We do so until $x_1 + x_2$ equals 1, or the common derivative equals $f'_3(0)$, whichever comes first. In the former case, we obtain the optimal solution (note that $x_i = 0$, $3 \leq i \leq n$). In the latter case, we need to introduce x_3. We can increase these three variables concurrently while keeping all three derivatives equal until the sum $x_1 + x_2 + x_3$ equals 1 or until the common derivative $f'_i(x_i)$, $1 \leq i \leq 3$, equals $f'_4(0)$, whichever comes first. We terminate the process when the sum of the active variables equals 1. Note that if x_n is utilized, we stop increasing the functions when $f_i'(x_i) = \lambda$, $1 \leq i \leq n$, for some constant λ and where $\Sigma_{i=1}^n x_i = 1$.

Remark. It is possible that the optimal solution for the program proposed in (3.1) is $x_1 = 1$ and $x_i = 0$, $2 \leq i \leq n$. This is the case when $f_1'(1) \leq f_i'(0)$, $2 \leq i \leq n$. Informally, for this we need the growth rate of the function $f_1(x)$ to be slower than the corresponding growth rates of all other functions. This will make the solution of the program rather trivial. In practice, it is quite rare to get such convex functions. However, suppose instead that the n functions $f_i(x)$, $1 \leq i \leq n$, are concave, namely $f_i''(x) < 0$, $1 \leq i \leq n$ (but they are still increasing). Here, the rate of growth is negative, so once $f_1(x_1)$ is set to a value higher than $x_1 = 0$, it is now more worthwhile to increase it further, in comparison with setting any of the values of other value to be larger than 0. In this case, the optimal solution is $x_1 = 1$ and $x_i = 0$, $2 \leq i \leq n$. In other words, the nature of the problem dictates that one of the variables dominates the others.

3.2.2 *Selecting the active variables*

The above procedure leads to the conclusion that the optimal solution possesses the following property: There exists some i, $1 \leq i \leq n$, such that $f_j'(x_j) = \lambda$, $1 \leq j \leq i$, for some constant λ, with $\Sigma_{i=1}^j x_i = 1$. Had the value of i been known in advance, the problem would have been reduced to solving a set of i non-linear equations with i variables, x_j, $1 \leq j \leq i$. Note that one of the equations is $\Sigma_{j=1}^i x_j = 1$. How do we find the value of i? The procedure in the previous section hints for the following algorithm. Try first $i = 1$. Clearly, $x_1 = 1$. Then, check whether or not $f_1'(1) \leq f_2'(0)$. If it does, the optimal solution has been reached. Otherwise, look for x_1 and x_2, such that $f'(x_1) = f'(x_2)$ and $x_1 + x_2 = 1$. Then, check whether or not $f'(x_1) < f_3'(0)$. In case it does, the optimal solution is obtained. Otherwise, introduce x_3, etc. Note that in the case where x_n is introduced, the optimal solution is when all $f_j'(x_j)$ are equal, $1 \leq j \leq n$. Of course, one still needs to mind the constraint $\Sigma_{j=1}^n x_j = 1$. This leads to a unique optimal solution.

3.2.3 *Example: minimizing the variance of the weighted average among independent random variables*

Let X_i be a random variable with variance of σ_i^2, $1 \leq i \leq n$. Assume that they are independent and share a common mean. In both sampling theory and in prediction, there is an interest in the following optimization problem:

$$\min_{w_1,\dots,w_n} \frac{1}{2}\mathrm{Var}\left(\sum_{i=1}^{n} w_i X_i\right),$$

$$\text{s.t.} \sum_{i=1}^{n} w_i = 1.$$

Indeed, the solution yields the minimum variance unbiased linear estimator for their common mean. The objective function is easily seen to equal $\frac{1}{2}\sum_{i=1}^{n} w_i^2 \sigma_i^2$. This is a separable objective function for which all single-variable functions are convex. Moreover, these single-variable functions share the same derivative of zero at zero. Hence, as argued in the previous section, the optimal solution is attained when all derivatives, namely $w_i \sigma_i^2$, $1 \le i \le n$, have the same value. This means that the optimal weights w_i, $1 \le i \le n$, are inversely proportional to the corresponding variance. Thus, for some constant C, $w_i = C/\sigma_i^2$, $1 \le i \le n$. The constant C is derived from the equality constraint that requires $\sum_{i=1}^{n} w_i = 1$. It is now straightforward that

$$w_i = \frac{\frac{1}{\sigma_i^2}}{\sum_{j=1}^{n} \frac{1}{\sigma_j^2}}, \quad 1 \le i \le n.$$

Note that the non-negativity constraint is redundant. Finally, it is easy to see that the optimal variance equals

$$\frac{1}{\sum_{i=1}^{n} \frac{1}{\sigma_i^2}},$$

namely it equals the harmonic mean of the n variances divided by n. This is better, by the definition of optimality, than $\frac{1}{n^2}\sum_{i=1}^{n} \sigma_i^2$ which is the corresponding arithmetic mean of the n variances divided by n, a value we would have gotten had we given equal weight to all n estimators. This fact exemplifies the inequality between various types of means (see (1.5)).

3.2.4 *Example: MLE for a multinomial distribution*

Suppose each individual in a population belongs to category k out of K categories with probability p_k, $1 \le k \le K$. Suppose a random sample of N individuals is taken. Let N_k be the random number of those out of these

N sampled who turned out to belong to category k. Then,

$$P(N_k = n_k, 1 \leq k \leq K) = \frac{N!}{\prod_{k=1}^{K} n_k!} \prod_{k=1}^{K} p_k^{n_k},$$

as long as $n_k \geq 0$ and $\Sigma_{k=1}^{K} n_k = N$. Note that the special case where $K = 2$ is the known binomial distribution. Suppose one is interested in the MLE for p_k, $1 \leq k \leq K$. Thus, one wishes to solve the following constrained optimization problem

$$\max_{p_k, 1 \leq k \leq K} \prod_{k=1}^{K} p_k^{n_k},$$

$$\text{s.t.} \quad \sum_{k=1}^{K} p_k = 1,$$

$$p_k \geq 0, \quad 1 \leq k \leq K,$$

or, equivalently, maximizing the logarithm of the above objective function

$$\max_{p_k, 1 \leq k \leq} \sum_{k=1}^{K} n_k \log p_k, \tag{3.2}$$

$$\text{s.t.} \quad \sum_{k=1}^{K} p_k = 1,$$

$$p_k \geq 0, \quad 1 \leq k \leq K.$$

Since $n_k \log p_k$ is a concave function whose derivative equals n_k/p_k, $1 \leq k \leq K$, and noticing that we face a maximization problem, then if $n_k \geq 1$, $1 \leq k \leq K$, we conclude that in the optimal solution

$$\frac{n_k}{p_k} = C, \quad 1 \leq k \leq K,$$

for some constant C, or equivalently,

$$p_k \propto n_k, \quad 1 \leq k \leq K.$$

The constraint that $\Sigma_{k=1}^{K} p_k = 1$, coupled with the fact that $\Sigma_{k=k}^{K} n_k = N$, implies that the MLE is

$$p_k = \frac{n_k}{N}, \quad 1 \leq k \leq K. \tag{3.3}$$

Finally, note that in the case where $n_k = 0$, this index does not appear in the objective function (3.2). Moreover, as this objective is monotone increasing with all other decision variables, it will be best to keep p_k at zero. Hence, the result stated in (3.3) holds also for the case where some of the resulting values are zero.

3.3 Queueing Examples

Under certain assumptions that we will not state here (for details see, e.g., [13, p. 61]), it is well-known that the mean queue length in a single-server queue with a Poisson arrival process with a rate of λ and exponential service times with mean μ^{-1}, where $\lambda < \mu$, equals

$$\frac{\lambda}{\mu - \lambda}. \tag{3.4}$$

This value includes the customer in service. Note that $\lambda/(\mu-\lambda)$ is a convex function both in λ and in μ. Moreover, it is monotone increasing in λ and monotone decreasing in μ. Also, $1/(\mu - \lambda)$ is the mean time in the system (service inclusive) for an individual customer. This too is a convex function in each of its two variables. In the next two subsections, we consider two optimization problems regarding multi-server queues.

3.3.1 *Optimal distribution of the service rate*

Suppose that there exists a set of n servers. The arrival rate at server i is known to be λ_i, $1 \leq i \leq n$. Suppose the total service rate of μ needs to be split among the n servers. For stability, it is required that $\mu > \Sigma_{i=1}^n \lambda_i$. Denote by x_i the decision variable of the service rate which is dedicated for server i, $1 \leq i \leq n$. The optimization problem of minimizing the sum of the n queue lengths is then

$$\min_{x_1,\ldots,x_n} \sum_{i=1}^n \frac{\lambda_i}{x_i - \lambda_i},$$

$$\text{s.t.} \ \sum_{i=1}^n x_i = \mu,$$

$$x_i > \lambda_i, \quad 1 \leq i \leq n.$$

See Figure 3.1(a).

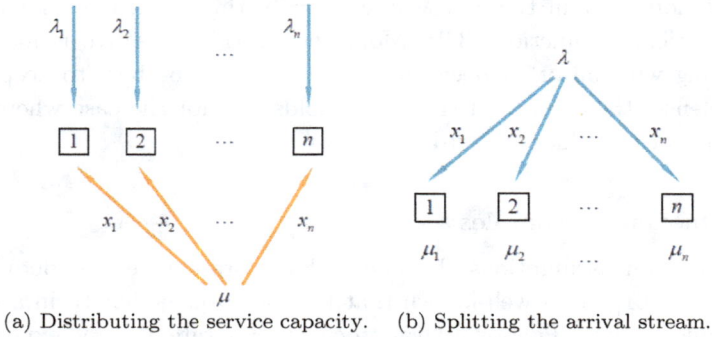

(a) Distributing the service capacity.   (b) Splitting the arrival stream.

Figure 3.1. Splitting service and arrival rates.

Clearly, $0 < x_i < \mu$, $1 \leq i \leq n$. Hence, the requirement for the equality of the derivatives in the optimal solution leads to

$$\frac{\lambda_i}{(x_i - \lambda_i)^2} = \alpha, \quad 1 \leq i \leq n,$$

for some α. From this it follows that

$$x_i = \lambda_i + \frac{\sqrt{\lambda_i}}{\sqrt{\alpha}}, \quad 1 \leq i \leq n. \tag{3.5}$$

By using the constraint $\Sigma_{i=1}^n x_i = \mu$,[a] we can solve for $\sqrt{\alpha}$ and get

$$\sqrt{\alpha} = \frac{\sum_{i=1}^n \sqrt{\lambda_i}}{\mu - \sum_{i=1}^n \lambda_i}.$$

Plugging this into (3.5), we conclude that

$$x_i = \lambda_i + \left(\mu - \sum_{j=1}^n \lambda_j\right) \frac{\sqrt{\lambda_i}}{\sum_{j=1}^n \sqrt{\lambda_j}}, \quad 1 \leq i \leq n. \tag{3.6}$$

This optimal allocation can be stated in words. First, the rate of service at server i should guarantee stability there, namely it needs to be greater than λ_i. This is the first term of λ_i, $1 \leq i \leq n$ in (3.6). Once this is taken care of, the next question is how to split the access service rate of $\mu - \Sigma_{i=1}^n \lambda_i$ among the n servers. The answer, given in (3.6), is that it needs to be done

---

[a] Had the constraint been $\Sigma_{i=1}^n x_i \leq \mu$, the optimal solution would still have satisfied $\Sigma_{i=1}^n x_i = \mu$.

proportionally to the square roots of the individual arrival rates. Simple algebra shows that the optimal objective function value equals

$$\frac{(\sum_{i=1}^{n} \sqrt{\lambda_i})^2}{\mu - \sum_{i=1}^{n} \lambda_i}. \tag{3.7}$$

Also, note that $-\alpha$ equals the derivative of the value of the optimal objective (3.7) with respect to μ. Note that this value captures the improvement of the objective function per an infinitesimal increase in the total service rate, currently at μ. As we will show in Section 4.4 below, this is by no means a coincidence. Finally, by inspecting (3.7), we conclude that as long as the optimal value is of concern, the $n + 1$ input data of n arrival rates and a single total service rate, are reduced to three entries: the sum of the arrival rates, the sum of their square roots, and the total service rate.

3.3.2   *Optimal splitting of the arrival rate*[b]

Suppose that there exists a set of n servers that face a joint arrival process at a rate of λ. The service rate at server i is known to be μ_i, $1 \leq i \leq n$. For stability, assume that $\sum_{i=1}^{n} \mu_i > \lambda$. The optimization problem dealt with here is how to split the arrivals among the n servers so as to minimize the total number of customers in the system. From (3.4), we conclude that this optimization problem can be formulated as

$$\min_{x_1,\ldots,x_n} \sum_{i=1}^{n} f_i(x_i),$$

$$\text{s.t. } \sum_{i=1}^{n} x_i = \lambda,$$

$$0 \leq x_i < \mu_i, \quad 1 \leq i \leq n,$$

where

$$f_i(x) = \frac{x}{\mu_i - x}, \quad 1 \leq i \leq n.$$

See Figure 3.1(b).

Note that $f_i'(x) = \frac{\mu_i}{(\mu_i - x)^2}$, $1 \leq i \leq n$. In particular, $f_i'(0) = 1/\mu_i$, which is the mean service time of a single customer at server i, $1 \leq i \leq n$. This is also the mean time of a customer in queue i, $1 \leq i \leq n$, assuming

---

[b]This problem appeared first in [4]. See also [12, pp. 63–64].

that nobody else seeks service there. Without loss of generality, assume that the servers are ordered from the fastest to the slowest, namely $\mu_1 \geq \mu_2 \geq \cdots \geq \mu_n$. Clearly, for a pair of servers i and j with $1 \leq i < j \leq n$, if $x_i = 0$ is part of the optimal solution, then of course $x_j = 0$. Also, note that $\lim_{x \to \mu_i} f_i(x) = \infty$, thus the upper bounds $x_i < \mu_i$, $1 \leq i \leq n$, are never binding in the optimal solution.

Suppose that servers 1 through i are open while the others are not, namely $x_j > 0$, $1 \leq j \leq i$ and $x_j = 0$, $i+1 \leq j \leq n$. Note that a lower bound on i is $\min_{1 \leq i \leq n}\{\Sigma_{j=1}^{i}\mu_j > \lambda\}$. Also, note that the possibility that $i = n$, namely all servers are open under the optimal solution, is not ruled out. Then, the common derivative condition argued for above is

$$f'(x_j) = \frac{\mu_j}{(\mu_j - x_j)^2} = \alpha_i, \quad 1 \leq j \leq i,$$

where α_i is the common derivative at the optimal arrival rates. Hence,

$$x_j = \mu_j - \frac{\sqrt{\mu_j}}{\sqrt{\alpha_i}}, \quad 1 \leq j \leq i,$$

This, coupled with the constraint $\Sigma_{j=1}^{i}x_j = \lambda$, leads to the solution

$$x_j = \mu_j - \frac{\sqrt{\mu_j}}{\sum_{k=1}^{i}\sqrt{\mu_k}}\left(\sum_{k=1}^{i}\mu_k - \lambda\right), \quad 1 \leq j \leq i, \qquad (3.8)$$

and

$$x_j = 0, \quad i+1 \leq j \leq n.$$

As it turns out,

$$\alpha_i = \frac{(\sum_{j=1}^{i}\sqrt{\mu_j})^2}{(\sum_{j=1}^{i}\mu_j - \lambda)^2}. \qquad (3.9)$$

For completeness, note that the value of the objective function at this solution equals

$$\sum_{j=1}^{i}\frac{x_j}{\mu_j - x_j} = \sum_{j=1}^{i}\frac{\mu_j - \frac{\sqrt{\mu_j}}{\sqrt{\alpha_i}}}{\frac{\sqrt{\mu_j}}{\sqrt{\alpha_i}}} = \sqrt{\alpha_i}\sum_{j=1}^{i}\sqrt{\mu_j} - i = \frac{(\sum_{j=1}^{i}\sqrt{\mu_j})^2}{\sum_{j=1}^{i}\mu_j - \lambda} - i.$$

$$(3.10)$$

Differentiate this value with respect to λ and note that you get α_i. Recall that this derivative measures (infinitesimally) how much the optimal objective value changes due to a change in the value of the input variable λ. The fact that this derivative equals α_i is not by chance. This sensitivity issue will be dealt with at length in Section 4.5 below. Finally, note that the derivative of the optimal objective value (3.10) with respect to μ_k equals

$$\alpha_i \left(\frac{\sum_{j=1}^{i} \mu_j - \lambda}{\sqrt{\mu_k} \sum_{j=1}^{i} \sqrt{\mu_j}} - 1 \right), \quad 1 \le k \le i.$$

This value needs to be negative as the objective value strictly improves when the service capacity of an open server increases. This point can be proved technically using the fact that $k \le i$. Note that this improvement is non-increasing in k, namely it is better to invest in improving the already fast servers. Note also that, trivially, the corresponding derivative with respect to μ_j, $i + 1 \le j \le n$, equals zero. The takeaway message is that it is not worthwhile at all to (locally) invest at the already very slow servers. Finally, deduce from (3.8), that

$$\frac{d\,x_j}{d\,\lambda} = \frac{\sqrt{\mu_j}}{\sum_{k=1}^{i} \sqrt{\mu_k}}, \quad 1 \le j \le i,$$

and

$$\frac{d\,x_j}{d\,\lambda} = 0, \quad i < j \le n.$$

From the above discussion, we learn that finding the optimal value boils down to finding the value of the optimal i. Using the argument described in Section 3.2.2, we conclude that this is the smallest value for i, $1 \le i \le n$, such that $\alpha_{i-1} > f_i'(0)$ and $\alpha_i \le f_{i+1}'(0)$. In other words, find the lowest value for i, such that $\alpha_{i-1} > 1/\mu_i$ and $\alpha_i \le 1/\mu_{i+1}$ (set μ_{n+1} to 0). Finding this value can be done by a brute force check: initialize with $i = 1$, check whether $\alpha_i < 1/\mu_{i+1}$. If this is the case, the optimal solution was just derived. In particular, $x_j = 0$, $i + 1 \le j \le n$. If not, increase the value of i by 1, etc. An alternative brute force approach is to compute the value stated in (3.10) for all i with $\Sigma_{j=1}^{i} \mu_j - \lambda > 0$, $1 \le i \le n$, and pick the one with the smallest value.

Exercise 3.1

The following problem appears in [1]. First, we define a queueing loss system. The arrival process and the service process are as defined in the queueing examples previously discussed, except now a customer that finds a busy server leaves the system forever. In the case where the arrival rate equals

λ and the service rate equals μ, the customers find server idle with a probability of $\mu/(\lambda + \mu)$. Second, assume n such servers exist, where server i serves with a rate of μ_i, $1 \le i \le n$. Third, assume a common arrival rate of λ. The question here is how to split the arrival stream among the n servers so as to maximize the expected number of successes per unit of time. Note that if x_i is the arrival rate to server i, then

$$x_i \cdot \frac{\mu_i}{x_i + \mu_i}$$

is the successes rate due to server i, $1 \le i \le n$. One wishes to maximize the sum of these values.

1. State formally the decision problem described above as a constrained optimization problem with one equality constraint.
2. Show that the objective function is concave in any of the n decision variables.
3. Argue why in this model all servers are open, namely at the optimal solution $x_i > 0$, $1 \le i \le n$.
4. Use the condition of equal derivatives in order to derive the optimal splitting of arrival rate.
5. What is the value of the optimal objective function?
6. What is the derivative of the value stated at the previous item with respect to λ?

3.3.3　*Who and how many should join a queue*

Consider a single server queue with a rate of service μ. Suppose there are n classes of customers, and a customer of class i suffers an amount of C_i per unit of time in the system and generates a value of R_i due to being granted service, $1 \le i \le n$. The optimization problem we face is what should be the arrival rate of each class? This is in fact (see (3.4)) the problem,

$$\max_{x_i, 1 \le i \le n} \sum_{i=1}^{n} \left(x_i R_i - C_i x_i \frac{1}{\mu - \sum_{j=1}^{n} x_j} \right),$$

$$\text{s.t.} \sum_{i=1}^{n} x_i \le \mu,$$

$$x_i \ge 0, \quad 1 \le i \le n.$$

We next solve this problem. The solution may surprise the reader.

First, consider the case where $n = 1$ and hence remove the reference to the index of the class. In this case, one maximizes the single-value function

$$f(x) = Rx - Cx\frac{1}{\mu - x}, \quad 0 \le x \le \mu.$$

This is a concave function, and by checking the FOCs, it is easy to see that maximization is attained at

$$x^* = \mu - \sqrt{\frac{C\mu}{R}}. \tag{3.11}$$

Moreover,

$$f(x^*) = (\sqrt{R\mu} - \sqrt{C})^2. \tag{3.12}$$

Second, and back to the general case, suppose that the total arrival rate is fixed to $\lambda < \mu$. Then, the problem we face is that

$$\max_{x_i, 1 \le i \le n} \left\{ \sum_{i=1}^n x_i \left(R_i - C_i \frac{1}{\mu - \lambda} \right) \right\},$$

$$\text{s.t.} \sum_{i=1}^n x_i = \lambda,$$

$$x_i \ge 0, \quad 1 \le i \le n.$$

The solution for this problem is immediate: Let

$$i^* = \arg \max_{i=1}^n \left\{ \lambda \left(R_i - C_i \frac{1}{\mu - \lambda} \right) \right\}.$$

Then, $x_{i^*} = \lambda$ and $x_i = 0$, $1 \le i \ne i^* \le n$. This is a typical *class-dominance* result: Those who are eventually admitted to service belong to a single class, while all others are refused entry. This fact implies also the optimal choice for λ.

Finally, combining this with (3.12), we conclude that the optimal value of the function equals

$$\max_{i=1}^n (\sqrt{R_i \mu} - \sqrt{C_i})^2,$$

where the optimal solution (see (3.11)) is $x_{i^*} = \mu - \sqrt{\frac{C_{i^*}\mu}{R_{i^*}}}$ and $x_i = 0$, $1 \le i \ne i^* \le n$, with

$$i^* = \arg \max_{i=1}^n (\sqrt{R_i \mu} - \sqrt{C_i})^2.$$

Note that the optimization problem we have solved is in fact a discrete one. In particular, it calls for an effort of $O(n)$ in order to be solved. For more details on this problem see [12, p. 57]. For many more queueing optimization problems see [22].

3.4 Optimization Under Linear Equality Constraints

3.4.1 *Main results*

For some function $f(x) : R^n \to R$, consider the following optimization problem:

$$\min_{x_1, x_2, x_3} f(x_1, x_2, x_3), \qquad (3.13)$$

$$\text{s.t. } x_1 + 2x_2 + 3x_3 = 5,$$

$$-x_1 + 3x_2 - x_3 = 4.$$

After putting in some thought, we can notice that fixing the value of one of the three decision variables determines, via the constraints, the values of the other two decision variables. This can be done by solving the resulting two-by-two system of linear equations. Thus, the objective function can be written in terms of a single variable, reducing the problem we face into a single-variable unconstrained optimization problem. In particular, all the machinery developed in the previous part of this text can be applied.

Adding one more constraint to the optimization problem stated in (3.13) leads to one, and only one, of the following three cases:

1. A regular 3×3 system of linear equations is formed. It will then have a unique solution, making the optimization part of the problem rather trivial.
2. The resulting three constraints are contradicting, namely a feasible solution does not exist. Specifically, suppose the added constraint is $ax_1 + bx_2 + cx_3 = d$. Then, a contradiction arises in the case where (a, b, c) is in the linear subspace spanned by $(1, 2, 3)$ and $(-1, 3, -1)$, namely there exist (unique) α_1 and α_2 such that $(a, b, c) = \alpha_1(1, 2, 3) + \alpha_2(-1, 3, -1)$. However, if $d \neq 5\alpha_1 + 4\alpha_2$, there is no feasible solution.
3. The third constraint is redundant. This is as in case (2) but with a key distinction: $d = 5\alpha_1 + 4\alpha_2$. In this case, any solution of the first two equations is also a solution of the third one.

The above argument suggests that the case of linear constraints is relatively easy, and we should move to the more general case. This is true. But first, lets go through the following analysis of the linear case as it sheds light on the general case where the constraints are not necessarily linear.

For some $f(x) : R^n \to R$, consider the following problem:

$$\min_{x \in R^n} \quad f(x),$$

$$\text{s.t. } Ax = b,$$

where $A \in R^{m \times n}$ is the *constraint matrix* and $b \in R^m$ is the *right-hand side vector*. Assume that $m < n$ and that A is a full-rank matrix. In particular, $rank(A) = m$. This assumption implies that there exists a square submatrix $B \in R^{m \times m}$ of A that is invertible. Without loss of generality, assume that B is the first m columns of A. Denote the corresponding subset of m decision variables by $x_B \in R^m$ and the rest of the decision variables by $x_N \in R^{n-m}$. Finally, denote by $N \in R^{m \times (n-m)}$ the rest of the columns of the constraint matrix. Thus, any feasible solution $x \in R^n$ satisfies

$$Bx_B + Nx_N = b.$$

This leads to $Bx_B = b - Nx_N$ and then to $x_B = B^{-1}(b - Nx_N)$. For example, if $x_N = \underline{0}$ then $x_B = B^{-1}b$. Thus, once the entries of x_N are determined, there is no more freedom in selecting x_B as it must obey the above equality. Put differently, the n-variable constrained optimization problem reduces to an $n - m$ unconstrained optimization problem. Finally, observe that the full-rank assumption implies that a feasible solution exists and that it is not unique. Note that uniqueness exists in the case where $m = n$. Note that had $m > n$, then the main issue would have been the existence of a feasible solution, and usually there are no such solutions.

Returning to the above example, suppose that B corresponds to the first two decision variables. Then,

$$\begin{pmatrix} x_1 \\ x_2 \end{pmatrix} = \begin{pmatrix} 1 & 2 \\ -1 & 3 \end{pmatrix}^{-1} \left[\begin{pmatrix} 5 \\ 4 \end{pmatrix} - x_3 \begin{pmatrix} 3 \\ -1 \end{pmatrix} \right].$$

In particular, both x_1 and x_2 can be expressed as affine functions of x_3. Plugging these two expressions into the original objective function, we can see that the original three-variable constrained optimization is now reduced to a single-variable unconstrained optimization, where the decision variable is x_3.

Let us reconsider the problem discussed above. Specifically,

$$f(x) = f(x_B, x_N) = f(B^{-1}(b - Nx_N), x_N) = F(x_N), \qquad (3.14)$$

for some function $F(y) : y^{n-m} \to R$. As said, we can now utilize all we did in Part I with respect to this function $F(y)$, yet we still want to look at the optimization problem from the point of view of the original function f. Specifically, focusing on the FOCs, we look for $x_N \in R^{n-m}$ such that $\nabla F(x_N) = \underline{0}$. Consider (3.14) and take the derivative with respect to x_N (using the chain rule (2.5)). Get that

$$\nabla F(x_N) = -N^T(B^{-1})^T \nabla_{x_B} f(B^{-1}(b - Nx_N), x_N)$$
$$+ \nabla_{x_N} f(B^{-1}(b - Nx_N), x_N).$$

Search for $x^* \in R^n$ that meets the FOCs $\nabla F(X_N^*) = \underline{0} \in R^{n-m}$, we get that for the vector $\lambda \in R^m$,

$$\lambda = (B^{-1})^T \nabla_{x_B} f(x^*),$$
$$-N^T \lambda + \nabla_{x_N} f(x^*) = \underline{0} \in R^{n-m}.$$

Moreover, by the definition of λ,

$$\nabla_{x_B} f(x^*) - B^T \lambda = \underline{0} \in R^m.$$

Combining the above two equality sets into one set, we get that for some $\lambda \in R^m$,

$$\nabla f(x^*) - A^T \lambda = \underline{0} \in R^n. \qquad (3.15)$$

To this, we have to add the original constraints, namely

$$Ax^* = b. \qquad (3.16)$$

In summary, note that the constrained optimization problem was reduced to a square system of (not necessarily linear) equations, (3.15) and (3.16), albeit with $n+m$ variables. In the (important) case where $f(x)$ is quadratic, the system of equations is in fact linear. More on this case will appear later.

Let us consider (3.15) and assume that $\nabla f(x^*)$ is given. Looking at $\lambda \in R^m$, we have a system with n linear equations and $m < n$ decision variables λ_i, $1 \le i \le m$. Such systems, as we have here more equations than variables, are usually infeasible, as the set of equations is overdetermined. Loosely speaking, a "miracle" seems to take place at x^*: The vector $\nabla f(x^*)$, which is in R^n, lies in the subspace spanned by the $m < n$ rows of the matrix A. Of course, this is not due to miracle but rather to the requirement that x^* is optimal, or at least, a stationary point.

Solving this set of equations, which we now know is feasible, we immediately get from (3.15) that

$$A\nabla f(x^*) - AA^T\lambda = \underline{0} \in R^m, \qquad (3.17)$$

and therefore,

$$\lambda = (AA^T)^{-1}A\nabla f(x^*). \qquad (3.18)$$

Note that the fact that $(AA^T)^{-1}$ exists follows from the assumption that A is a full-rank matrix. See [14, p. 71]. Plugging this value of λ into (3.15), we can conclude that a necessary condition for a feasible solution x^* to be a stationary point is that

$$\nabla f(x^*) = A^T(AA^T)^{-1}A\nabla f(x^*). \qquad (3.19)$$

Let us put the above analysis into words. As already noted, a necessary condition for a feasible solution $x^* \in R^n$ to be a stationary constrained solution is that the gradient of the objective function at this point belongs to the m-dimensional linear subspace spanned by the m rows of A. Taking this a bit further, we can say that the m rows are the gradients of the m equality constraints. Thus, a necessary condition for $x^* \in R^n$ to be a stationary constrained solution is that the gradient of the objective function at this point belongs to the m-dimensional linear subspace spanned by the m gradients of the constraints. This final statement (rightly) looks overreaching in the case of linear constraints, but as we will see below, it is exactly the relevant condition when the constraints are not necessarily linear. Finally, the m coefficient determining the linear combinations are λ_i, $1 \leq i \leq$, and they are referred to as the *Lagrange multipliers*.

Remark. Let us consider (3.15) again. It can be viewed as a linear system of equations with the unknown variables $\lambda \in R^m$. This is a set of n equations with m unknown variables. Since $m < n$, this is an overdetermined system. Such systems usually do not have a solution, but the assumptions made here indicate that we are facing the rare case where a solution does exist. Moreover, by our assumption of a full-rank matrix, the solution is unique, and hence it is given in (3.18). Since $A^T(AA^T)^{-1}Ax = b$ for some $b \in R^n$ only if $A^Ty = b$ for some $y \in R^m$ ($y = (AA^T)^{-1}Ax$, of course), we can conclude that $rank(A^T(AA^T)^{-1}A) = m$. Equivalently, zero is an eigenvalue of the symmetric matrix $A^T(AA^T)^{-1}A$ with a (algebraic and geometric) multiplicity of $n - m$. Moreover, since $A^T(AA^T)^{-1}Ax = \alpha x$ for some scalar α only if $Ax = \alpha Ax$, we can conclude that the only non-zero eigenvalue of $A^T(AA^T)^{-1}A \in R^{n \times n}$ is 1 with a (algebraic and geometric) multiplicity of m. From (3.19) it follows that a necessary condition for x^* to be a stationary point is that $\nabla f(x^*)$ is an eigenvector of $A^T(AA^T)^{-1}A$ corresponding to the eigenvalue of 1. Put differently, the search for x^* is limited to points x where $\nabla f(x)$ belongs to this eigenspace.

Remark. An alternative argument for (3.19) is the following. By (2.15), we have that $A^T(AA^T)^{-1}A$ is the projection operation on the linear subspace spanned by the rows of A. In particular, it is the identity operation among vectors belonging to this subspace. Finally, from (3.15) we infer that $\nabla f(x^*)$ belongs to this subspace that has a dimension of m.

3.4.2 *Example: linearly constrained minimum norm*

For $B \in R^{n \times n}$, $c \in R^n$, $b \in R^m$, and $A^{m \times n}$ with $m \leq n$, where B is invertible and $rank(A) = m$, consider the following problem:

$$\min_{x \in R^n} \frac{1}{2}\|Bx - c\|^2,$$

$$\text{s.t. } Ax = b.$$

Note that one can assume without loss of generality that $B = I$ and $c = \underline{0}$. Otherwise, change the variables by replacing x with $y = Bx - c$, and $Ax = b$ with $AB^{-1}y = b - AB^{-1}c$. Once this problem is solved for y^*, the solution to the original problem is $x^* = B^{-1}(y^* + c)$. Thus, the problem is

$$\min_{y \in R^n} \frac{1}{2}\|y\|^2, \tag{3.20}$$

$$\text{s.t. } Ay = b.$$

This problem is usually referred to as the minimum-norm problem. It can be solved explicitly. Specifically, noticing that now $\nabla f(y) = y$, and using (3.19), we get that

$$y^* = A^T(AA^T)^{-1}Ay^* = A^T(AA^T)^{-1}b, \tag{3.21}$$

which is the solution we are after. Using the notation introduced in (2.16), we can say that $y^* = (A^T)^\dagger b$. The fact that this is the unique global minimum point is argued in the next example.

We are interested in a special case of the above problem, namely

$$\min_{x \in R^n} ||x - c||^2,$$

$$\text{s.t. } a^T x = b,$$

where $a, c \in R^n$, and $b \in R$. The set of points $x \in R^n$ that satisfy $a^T x = b$ is called a *hyperplane*. Thus, we want to identify the closest point in this hyperplane to some given point c. Based on (3.21) and the change of scale where the scalar b is replaced with $b - a^T c$, we can see that this point is

$$x^* = \frac{b - a^T c}{||a||^2} a + c,$$

while the the distance itself equals

$$|b - a^T c|/||a||. \tag{3.22}$$

This distance is referred to as the (orthogonal) distance between c and the hyperplane.

A specific case of this is when we consider the distance between two parallel hyperplanes. Specifically, two hyperplanes $a_i^T x = b_i$, $i = 1, 2$, are said to be parallel if $a_1 = a_2$, which is accordingly denoted by a. From (3.22), we infer that in this case the distance between the two hyperplanes equals $|b_1 - b_2|/||a||$. An alternative argument is as follows. Suppose for $x_0 \in R^n$, $a^T x_0 = b_1$. Then, consider the following optimization problem:

$$\min_{x \in R^n} ||x - x_0||^2,$$

$$\text{s.t. } a^T x = b_2.$$

Finally, the distance itself equals

$$|b_2 - a^T x_0|/||a|| = |b_2 - b_1|/||a||. \tag{3.23}$$

Note that, as expected, the choice of x_0 does not matter (as long as it is on the first hyperplane).

3.4.3 *Example: linearly constrained least squares*

The unconstrained least squares problem has been discussed in Section 2.2.1. A similar objective, but with the added linear equality constraints, is

$$\min_{x \in R^n} \frac{1}{2} ||Ax - b||^2,$$

$$\text{s.t. } Cx = d,$$

for a full-rank matrix $A \in R^{m \times n}$ with $m \geq n$ and some full-rank matrix $C \in R^{p \times n}$ for some $p \leq n$. Observe that the assumption where $m \leq n$ is the key distinction between the problem dealt with here and the one dealt with in the previous example. Admittedly, the analysis done below holds also for the case where $m = n$, making the previous example a special case of the current one.

Note that the full-rank assumption implies that the rows of C are linearly independent. Observe that $\nabla f(x) = A^T Ax - A^T b$. Now (3.15), coupled with the equality constraints can be written as

$$\begin{pmatrix} A^T A & -C^T \\ C & \underline{0} \end{pmatrix} \begin{pmatrix} x \\ \lambda \end{pmatrix} = \begin{pmatrix} A^T b \\ d \end{pmatrix}. \tag{3.24}$$

Our next aim is to prove the existence and the uniqueness of the stationary point. This is equivalent to arguing that

$$\begin{pmatrix} A^T A & -C^T \\ C & \underline{0} \end{pmatrix} \tag{3.25}$$

is invertible. Following [8, pp. 345–346], we prove the invertability by showing that if

$$\begin{pmatrix} A^T A & -C^T \\ C & \underline{0} \end{pmatrix} \begin{pmatrix} x \\ \lambda \end{pmatrix} = \begin{pmatrix} \underline{0} \\ \underline{0} \end{pmatrix}, \tag{3.26}$$

then $x = \underline{0}$ and $\lambda = \underline{0}$. Indeed, the above is equivalent to $A^T Ax - C^T \lambda = \underline{0}$ and $Cx = \underline{0}$. Multiplying from the left the first set of equations by x^T, we get that $||Ax||^2 - x^T C^T \lambda = 0$. But the second set of equations says that $x^T C^T = \underline{0}^T$, and so we conclude that $||Ax||^2 = 0$, which, by the full-rank assumption on A, implies that $x = \underline{0}$. Finally, since $Ax = \underline{0}$, the first set of equations, $x^T Ax - C^T \lambda = \underline{0}$, is in fact $-C^T \lambda = \underline{0}$. The full-rank assumption on C implies that $\lambda = \underline{0}$, as required. This completes the proof.

Finally, we claim that the resulting stationary point is a minimum point. This is based on the fact that $F(x_N)$ (as defined in (3.14)) is convex in x_N (as implied immediately by the fact that $f(x) = ||Ax - b||^2$ is convex and by Theorem 2.8) (see also the example following this theorem). We revisit this issue later when Lagrange multipliers and convex optimization are introduced. For an alternative proof see [8, pp. 346–347].

3.4.4 *Example: linear objective function*

Suppose that $f(x) = c^T x$ for some $c \in R^n$. Then, to use the notation of (3.14), if we write c^T as $c^T = (c_B^T, c_N^T)$ and then $c^T x$ as $c_B^T x_B + p_N^T x_N$, we get that

$$f(x) = c_B^T B^{-1} b + (c_N^T - p^T B^{-1} N) x_N,$$

which is a linear function in x_N. Thus, unless $c_N^T - c^T B^{-1} N = \underline{0}^T$, where $f(x) = c_B^T B^{-1} b$ for any $x \in R^n$, this function can be made as small or as large as one wishes. Thus, unless one adds some (linear) inequality constraints, for example requiring that all variables are non-negative, such optimization problems are rather trivial. Once such inequalities are introduced, a vast set of problems and algorithms is opened and leads to a dedicated branch of optimization of its own right, that of *linear programming*. For more, see Chapters 6–8 below.

Chapter 4

Optimization Under Equality Constraints: The General Case

4.1 An Informal Introduction to First-Order Conditions (FOCs)

We start with a reminder. Let $f(x)$ be a function from R^n to R. A point $x \in R^n$ is called a *stationary* point if $\nabla f(x) = \underline{0} \in R^n$. A necessary condition for a point x_0 to be a local extreme point, be it a minimum or maximum point, is that it is a stationary point. The reason for this is as follows: Without loss of generality, assume that $x_0 = \underline{0}$ and $f(\underline{0}) = 0$. Then, the linear approximations for $f(x)$ and $f(-x)$ around $x_0 = \underline{0}$ (see (2.2)) are

$$f(x) \approx \nabla f(\underline{0})^T x \quad \text{and} \quad f(-x) \approx -\nabla f(\underline{0})^T x,$$

respectively. Thus, unless $\nabla f(\underline{0})^T x = 0$ for all $x \in R^n$, which is possible if and only if $\nabla f(x) = \underline{0}$, namely if $\underline{0}$ is a stationary point, the function $f(x)$ cannot have an extreme point at $\underline{0}$. This is the case as otherwise, one of the above two values would have been positive, while the other would have been negative. In particular, $\underline{0}$ would neither be a local maximum nor a local minimum point. As said, $\nabla f(\underline{0})^T x = 0$ for all $x \in R^n$ if and only if $\nabla f(\underline{0}) = \underline{0}$. Indeed, otherwise, namely if $\nabla f(\underline{0}) \neq \underline{0}$, a counterexample would be $x = \epsilon \nabla f(\underline{0})$ for some small $\epsilon > 0$, as then $\epsilon \nabla f(\underline{0})^T \nabla f(\underline{0}) > 0$.

Suppose now that one wishes to find an extreme point for $f(x)$ but under the constraint that $g(x) = 0$, where $g(x)$ is also a function from R^n to R. Note that the zero on the right-hand side is without loss of generality. Suppose that x_0 is a feasible point, namely $g(x_0) = 0$. Hence, the issue is whether or not this is a (feasible) local extreme point of $f(x)$. The answer is "yes" if there exists an $\epsilon > 0$ such that for any x with $g(x) = 0$ and with $||x - x_0|| < \epsilon$, $f(x_0) \leq f(x)$. Among all directions which emanate from

x_0, only those that (locally) preserve the constraint are of interest. Indeed, what happens along directions that (locally) take us outside the feasible set is immaterial.

A first-order approximation for $g(x)$ around x_0 is $g(x_0) + \nabla g(x_0)^T$ $(x - x_0) = \nabla g(x_0)^T(x - x_0)$. Note that this approximation is error-free in the case where $g(x)$ is affine. Thus, if we ignore second-order terms, we are interested in x, or more precisely, in the direction $x - x_0$, only in the case where $\nabla g(x_0)^T(x - x_0) = 0$, as other directions take us outside the feasible set. Considering $x - x_0$ as a direction denoted by d, where a tiny step in this direction does not (practically) violate the constraint. For that to be the case, we are interested in directions d that are orthogonal to $\nabla g(x_0)$, namely $\nabla g(x_0)^T d = 0$. We call such directions *feasible directions* at x_0, since starting at x_0 and moving infinitesimally along these directions (whether backwards or forwards) does not, up to a second-order term, violate the constraint.

Note that the set of feasible directions forms a linear subspace. In particular, if d is such a direction, then so is the case with $-d$. Other directions are of no interest: In order to get a local constrained extreme point for $f(x)$ at x_0, we need x_0 to be a stationary point for $f(x)$ but only along some directions, directions that preserve the local feasibility of x_0. Moreover, for optimality with respect to the function $f(x)$, we need $\nabla f(x_0)$ to also be orthogonal to all these directions. This means that $\nabla f(x_0)$ and $\nabla g(x_0)$ need to be with the same (up to a sign) direction, or equivalently, $\nabla f(x_0)$ needs to be in the linear subspace spanned by $\nabla g(x_0)$.

Suppose now that there are two constraints, $g_1(x) = g_2(x) = 0$. Now, the set of directions of interest is those that lie in the intersection of the two sets when each of these two constraints is treated individually, as done above. Note that the intersection between two linear subspaces is a linear subspace too, but of a reduced dimension (see, e.g., [14, p. 164]). For the latter fact to be strictly true, we make the technical assumption (called *regularity*) that $\nabla g_1(x_0)$ and $\nabla g_2(x_0)$ are linearly independent, namely one is not a constant multiplier of the other. Now the condition for a locally constrained extreme point is that $\nabla f(x_0)$ is orthogonal to all the directions that are orthogonal to both $\nabla g_1(x_0)$ and $\nabla g_2(x_0)$. In summary, if for $d \in R^n$, $\nabla g_i(x_0)^T d = 0$, $i = 1, 2$, then $\nabla f(x_0)^T d = 0$ too. Some imagination will convince the reader why the set of vectors $d \in R^n$, with $\nabla g_i(x_0)^T d$, $i = 1, 2$, for a feasible point x_0 is called the *tangent plane* (for the feasible set) at x_0. This means that $\nabla f(x_0)$ belongs to the subspace that is formed by all vectors that are orthogonal to all those which are orthogonal to both

$\nabla g(x_1)$ and $\nabla g(x_2)$. Hence, it belongs to the subspace spanned by these two vectors. This is equivalent to requiring the existence of two scalars λ_1 and λ_2, such that

$$\nabla f(x_0) = \lambda_1 \nabla g_1(x_0) + \lambda_2 \nabla g_2(x_0).$$

Moreover, when these scalar exist, they are unique. In the optimization terminology, λ_1 and λ_2 are called *Lagrange multipliers* or *dual variables*.

The above argument can be extended to the case of m constraints as long as $m \leq n$. The requirement that $m \leq n$ enables us to assume that the m gradients $\nabla g_i(x_0)$, $1 \leq i \leq m$, that are of course vectors in R^n, are linearly independent. Thus, the requirement for stationarity at x_0, is the existence of (unique) scalars $\lambda_1, \lambda_2, \ldots, \lambda_m$ such that

$$\nabla f(x_0) = \sum_{i=1}^{m} \lambda_i \nabla g_i(x_0). \tag{4.1}$$

On the one hand, note that the larger m is, the harder it is to have $g_i(x_0) = 0$, $1 \leq i \leq m$, as a larger set of constraints is concerned. On the other hand, once a point is feasible, then a larger m means that the larger the subspace spanned by the gradients of the constraints is, making the existence of a set of Lagrange multipliers more likely. Put differently, on the face of it, it seems that the search for a larger number of Lagrange multipliers makes success less likely. In fact, the opposite is the case. As a sanity check, note that in the case where $m = n$ and the regularity condition holds, feasibility is sufficient for optimality: The gradients of the constraints span R^n and trivially $\nabla f(x_0)$ lies there. Later on we will show an example where the gradients of the constraints are not linearly independent at a point known to be optimal, namely the regularity condition does not hold, leading to the non-existence of Lagrange multipliers.

Remark. We consider above stationary points, but usually our interest is minimum or maximum points. The above discussion shows that, at least from the point of view of Lagrange multipliers, both type of points meet the same conditions, and hence something further is required in order to distinguish between these two extremes.

Remark. It is true that stationarity does not necessarily lead to optimality. However, when solving an optimization problem using the above approach, the problem of optimizing under equality constraints is reduced to a problem of solving a set of equations with $m + n$ variables. True, these equations are usually not linear, but as we will see through a number of examples, they can sometimes be solved.

4.2   First-Order Conditions (FOCs): Theory and Examples

Recall the definition for a stationary point x for the function $f : R^n \to R$: $\nabla f(x) = \underline{0}$. The point of departure was the requirement that $\nabla f(x)^T d = 0$ for any direction vector $d \in R^n$ (which is met if and only if $\nabla f(x) = \underline{0}$). When we switched to optimization under constraints, this requirement is relaxed to only feasible directions, namely for directions $d \in R^n$ that satisfy $\nabla g_i(x)^T d = 0$, $1 \leq i \leq m$. An equivalent condition is stated in Theorem 4.1 below. But first, we need the following definition.

Definition

Definition 4.1. A feasible point is said to be regular if the m vectors $\nabla g_i(x^*) \in R^n$, $1 \leq i \leq m$, are linearly independent.

Theorem

Theorem 4.1. *Consider the following optimization problem:*

$$\min_{x \in R^n} f(x),$$

$$s.t. \quad g_i(x) = 0, \quad 1 \leq i \leq m,$$

for some $m \leq n$. *Assume* x^* *to be a feasible point, namely* $g_i(x^*) = 0$, $1 \leq i \leq m$. *Also, assume that* x^* *is regular. Then, a necessary condition for the stationarity of* x^* *is that there exists a vector* $\lambda \in R^m$ *such that*

$$\nabla f(x^*) - \sum_{i=1}^{m} \lambda_i \nabla g_i(x^*) = \underline{0} \in R^n. \tag{4.2}$$

In words, $\nabla f(x^*)$ *lies in the linear subspace spanned by* $\nabla g_i(x^*)$, $1 \leq i \leq m$.

Proof. An informal explanation of this theorem appears in the previous section. A rigorous proof appears in Section 4.2.2.   □

The vector $\lambda \in R^m$ is usually referred to as the *Lagrange multipliers* or *dual variables*. Taken together, the feasibility conditions and the conditions imposed in (4.2) are known as the *first-order conditions* (FOCs) or the Karush-Kuhn-Tacker (KKT) conditions, for optimality.

Definition

Definition 4.2. Define the Jacobian matrix $\nabla \underline{g}(x) \in R^{m \times n}$ to be

$$[\nabla \underline{g}(x)]_{ij} = \frac{dg_i(x)}{dx_i}, \quad 1 \le i \le m, \quad 1 \le j \le n.$$

In other words, the transpose of gradient of $g_i(x)$ is the ith row of $\nabla \underline{g}(x)$.

In a similar fashion as (3.19) was derived, it is possible to observe that

$$\nabla f(x^*) = \nabla \underline{g}(x^*)^T [\nabla \underline{g}(x^*) \nabla \underline{g}(x^*)^T]^{-1} \nabla \underline{g}(x^*) \nabla f(x^*).$$

This reinforces the fact that the analysis of the general case just follows that of the linearly constrained case, where the original constraints have been linearized around x^*.

Remark. The regularity conditions assumed in Theorem 4.1 implies that if the Lagrange multipliers exist, then they are unique. It is tempting to conclude that if these conditions are not met, a necessary condition for optimality is that these multipliers exists but are not necessarily unique. However, this conjecture is false. As Example 8 in the next section demonstrates, it is possible that the multipliers do not exist at all, in spite of the optimality of x^*, and this is due to the fact that the regularity conditions do not hold.

Definition

Definition 4.3. The function $L(x, \lambda) : R^{n+m} \to R$ where

$$L(x, \lambda) = f(x) - \lambda^T \nabla \underline{g}(x) = f(x) - \sum_{i=1}^{m} \lambda_i g_i(x)$$

is called the *Lagrangian function*.

The first-order conditions, coupled with the feasibility conditions, can be put in a concise form as $\nabla L(x, \lambda) = \underline{0} \in R^{n+m}$. Moreover, by (4.2), we can see that this part of the FOCs, for a given $\lambda \in R^m$, deals with the unconstrained optimization problem

$$optimize_{x \in R^n} \left\{ f(x) - \sum_{i=1}^{m} \lambda_i g_i(x) \right\}.$$

Put differently, the FOCs for optimality in the case of equality constraints imply that if somehow, miraculously, the vector of m Lagrange multipliers is known, then solving a constrained optimization reduces to solving an unconstrained optimization problem.

4.2.1 *Examples*

A number of problems involving optimization under equality constraints are stated below. For all of them, using the FOCs defined above, we derive what turns out to be the unique stationary point coupled with the corresponding Lagrange multipliers. We claim, at this stage without a proof, that they indeed solve the stated optimization problem, be it one of minimization or maximization.

Example 1. Consider the following constrained optimization:

$$\min_{x_1, x_2, x_3} \frac{1}{2}(x_1^2 + x_2^2 + x_3^2),$$

$$\text{s.t.} \quad x_1 + x_2 = 1,$$

$$x_2 + x_3 = 1.$$

We next prove that $(1/3, 2/3, 1/3)$ is a stationary point. Firstly, it is easy to check that this is a feasible point. Secondly, the gradient of the objective function at this point is $(1/3, 2/3, 1/3)$. Thirdly, the gradient of the first constraint at this point (or, in fact, at any other point) is $(1, 1, 0)$, and the gradient of the second constraint is $(0, 1, 1)$. Note that these two gradients are linearly independent. Finally, note that

$$\begin{pmatrix} \frac{1}{3} \\ \frac{2}{3} \\ \frac{1}{3} \end{pmatrix} = \frac{1}{3} \begin{pmatrix} 1 \\ 1 \\ 0 \end{pmatrix} + \frac{1}{3} \begin{pmatrix} 0 \\ 1 \\ 1 \end{pmatrix}.$$

In other words, $\lambda_1 = 1/3$ and $\lambda_2 = 1/3$ are the unique Lagrange multipliers. Note that the key point here is not their uniqueness but their existence.

It is possible to see that this example is in fact a numerical version for Example 1 in Section 3.4 that deals with minimizing the norm of a vector under linear equality constraints. In particular, it is an example of (3.20). Specifically,

$$A^T(AA^T)^{-1}A = \frac{1}{3}\begin{pmatrix} 2 & 1 & -1 \\ 1 & 2 & 1 \\ -1 & 1 & 2 \end{pmatrix},$$

and so

$$\begin{pmatrix} \frac{1}{3} \\ \frac{2}{3} \\ \frac{1}{3} \end{pmatrix} = \frac{1}{3}\begin{pmatrix} 2 & 1 & -1 \\ 1 & 2 & 1 \\ -1 & 1 & 2 \end{pmatrix}\begin{pmatrix} \frac{1}{3} \\ \frac{2}{3} \\ \frac{1}{3} \end{pmatrix}.$$

As it turns out, the vector λ defined here is the same as the one that is defined in (3.15). Finally, by (3.15) and (3.16), we get that

$$\begin{pmatrix} 1 & 0 & 0 & -1 & 0 \\ 0 & 1 & 0 & -1 & -1 \\ 0 & 0 & 1 & 0 & -1 \\ 1 & 1 & 0 & 0 & 0 \\ 0 & 1 & 1 & 0 & 0 \end{pmatrix}\begin{pmatrix} \frac{1}{3} \\ \frac{2}{3} \\ \frac{1}{3} \\ -\frac{1}{3} \\ \frac{1}{3} \end{pmatrix} = \begin{pmatrix} 0 \\ 0 \\ 0 \\ 1 \\ 1 \end{pmatrix}.$$

Example 2 (Least squares under linear equality constraints). This (important) problem was already stated as Example 2 in Section 3.4. It is possible to show that the variables that were defined in (3.24) are the stationary point and the corresponding Lagrange multipliers. We leave this proof to the reader. See also [8, pp. 344–346].

Example 3 (The Cobb–Douglas model). Consider the following optimization problem:

$$\max_{x_1, x_2} x_1^a x_2^b,$$

$$\text{s.t. } p_1 x_1 + p_2 x_2 = m.$$

Note that one can assume, without loss of generality, that $m = 1$. We next argue that the point (x_1, x_2) where

$$x_1 = \frac{a}{a+b} \cdot \frac{m}{p_1} \quad \text{and} \quad x_2 = \frac{b}{a+b} \cdot \frac{m}{p_2}$$

meets the FOCs . Firstly, note that the Lagrangian function is

$$x_1^a x_2^b - \lambda(p_1 x_1 + p_2 x_2 - m).$$

Taking the derivatives with respect to x_1, x_2 and λ and setting them to zero yields

$$a x_1^{a-1} x_2^b - \lambda p_1 = 0,$$

$$b x_1^a x_2^{b-1} - \lambda p_2 = 0,$$

$$p_1 x_1 + p_2 x_2 - m = 0.$$

Note that the third equation is the budget constraint in disguise. We next look for a solution with $\lambda \neq 0$. Multiplying the first equation by bx_1 and the second by ax_2 leads to $p_1 b x_1 = p_2 a x_2$. Plugging one of the variable as a function of the other into the third equation (namely the constraint) completes the proof. Note that an alternative way of reaching the same conclusion is to apply the argument used in Section 3.2. Indeed, the problem we solved here is an example of an allocation problem. Finally, note that a solution with $\lambda = 0$ does not exist.

Example 4 (Distributing the service capacity (version 1)). We revisit the problem of the optimal distribution of the total service capacity among various servers in the same queueing system introduced and solved in Section 3.3.1. Let us ignore, for the time being, the constraints $x_i > \lambda_i$, $1 \leq i \leq n$. Using now α for the single Lagrange multiplier, we get that

$$L(x_1, \ldots, x_n, \alpha) = \sum_{i=1}^n \frac{\lambda_i}{x_i - \lambda_i} - \alpha \left(\sum_{i=1}^n x_i - \mu \right).$$

Taking the derivative with respect to x_i and setting it to zero yields

$$-\frac{\lambda_i}{(x_i - \lambda_i)^2} - \alpha = 0, \quad 1 \leq i \leq n.$$

From here on, the analysis repeats what was done in Section 3.3.1 and therefore is omitted. For future reference, note that the Lagrange multiplier equals

$$\alpha = -\frac{(\sum_{i=1}^n \sqrt{\lambda_i})^2}{(\mu - \sum_{i=1}^n \lambda_i)^2}. \text{[a]} \tag{4.3}$$

---

[a] The multiplier α here equals $-\alpha$ as the latter was defined in Section 3.3.1.

Note that the obtained solution satisfies $x_i > \lambda_i$, $1 \leq i \leq n$, keeping it a candidate for the optimal solution. Less formally put, we can say that this set of conditions plays no role when we consider the vicinity of the point we have reached.

Example 5 (Distributing the service capacity (version 2)). Suppose, as above, that there are n servers, where the arrival rate to server i equals λ_i, $1 \leq i \leq n$. However, here we assume that a unit of service rate at server i costs $p_i > 0$, $1 \leq i \leq n$. The problem now is to find the optimal service rates purchased so as to minimize the total cost among all those rates that meet the constraint that the (mean) sum of the queue lengths is bounded by some number L. The resulting problem is

$$\min_{x_i, 1 \leq i \leq n} \sum_{i=1}^{n} p_i x_i,$$

$$\text{s.t.} \sum_{i=1}^{n} \frac{\lambda_i}{x_i - \lambda_i} \leq L,$$

$$x_i > \lambda_i, \quad 1 \leq i \leq n$$

Clearly, in the optimal solution the constraint is binding. This implies that the inequality constraint can be assumed to be an equality constraint. On the other hand, all capacity constraints are non-binding so they will be ignored for the moment. Then, the Lagrangian function is

$$L(x_1, \ldots, x_n, \alpha) = \sum_{i=1}^{n} p_i x_i - \alpha \left(\sum_{i=1}^{n} \frac{\lambda_i}{x_i - \lambda_i} - L \right),$$

and its derivative with respect to x_i equals

$$p_i - \alpha \frac{\lambda_i}{(x_i - \lambda_i)^2}, \quad 1 \leq i \leq n.$$

Setting these derivatives to zero and solving for x_i as a function of α yields

$$x_i = \lambda_i + \sqrt{\alpha} \sqrt{\frac{\lambda_i}{p_i}}, \quad 1 \leq i \leq n.$$

Using the equality constraint and solving for $\sqrt{\alpha}$ implies that

$$\sqrt{\alpha} = \frac{\sum_{j=1}^{n} \sqrt{\lambda_j p_j}}{L}.$$

Finally, the optimal rates are

$$x_i = \lambda_i + \frac{\sum_{j=1}^{n}\sqrt{\lambda_j p_j}}{L}\sqrt{\frac{\lambda_i}{p_i}}, \quad 1 \le i \le n.$$

Note that these rate meet the n capacity constraints. Inserting these values for the decision variables in the objective function, leads to an optimal total cost which equals

$$\sum_{i=1}^{n}\lambda_i p_i + \frac{1}{L}\left(\sum_{i=1}^{n}\sqrt{\lambda_i p_i}\right)^2.$$

Interestingly, we got that the optimal cost is an affine function of $1/L$, with $(\Sigma_{i=1}^{n}\sqrt{\lambda_i p_i})^2$ being its slope and $\Sigma_{i=1}^{n}\lambda_i p_i$ its intercept. In particular, with respect to the optimal objective function value, the $2n$-size data of n arrival rates and of n prices per service rate reduce to only two entries, $\Sigma_{i=1}^{n}\lambda_i p_i$ and $\Sigma_{i=1}^{n}\sqrt{\lambda_i p_i}$.

Example 6 (Eigensystems of symmetric matrices). This example is a fundamental one, and hence it is upgraded below to a theorem. Let us recall that for a matrix $A \in R^{n\times n}$, a scalar λ and a non-zero vector $x \in R^n$ are called a *right eigenpair* of an eigenvalue and eigenvector, respectively, if $Ax = \lambda x$. A left eigenpair is defined accordingly. Note that for a symmetric matrix, (λ, x) is a right eigenpair if and only if (λ, x^T) is a left eigenpair. Also recall that orthogonal vectors are also linearly independent. See, e.g., [14, p.29]. The following theorem deals with the existence of eigenpairs in the case of symmetric matrices. This theorem does not extend to asymmetric matrices.

Theorem

Theorem 4.2. *For a symmetric matrix $A \in R^{n\times n}$, there exists an orthonormal basis for R^n that is formed of n eigenvectors.*

Proof. Consider the following optimization problem.

$$x_1 \in \arg\max_{x \in R^n} x^T A x,$$

$$\text{s.t. } ||x||^2 = 1.$$

The existence of $x_1 \neq \underline{0}$ is not an issue: One looks for a maximization over a compact set. This solution then needs to satisfy the FOCs. For a scalar λ, let

$$L(x, \lambda) = x^T A x - \lambda(x^T I x - 1).$$

Then,

$$\nabla_x L(x, \lambda) = 2Ax - 2\lambda x.$$

Therefore, for some λ_1, $Ax_1 = \lambda_1 x_1$. In particular, λ_1 and x_1 form an eigenpair. Note that the corresponding objective function value equals $x_1^T A x_1 = \lambda_1 x_1^T x_1 = \lambda_1$. Also, it cannot be ruled out that x_1 is not unique. In this case, x_1 is selected arbitrarily from the set of optimizer vectors. Next, for i with $2 \leq i \leq n$, define recursively the ith optimization problem to be

$$x_i \in \arg \max_{x \in R^n} x^T A x,$$

$$\text{s.t. } ||x||^2 = 1,$$

$$x_j^T x = 0, \quad 1 \leq j \leq i - 1.$$

Note that x_i is orthogonal to all vectors that solve the previous $i - 1$ optimization problems, of which the first one x_1 was already defined above. Again, the existence of $x_i \neq 0$, $1 \leq i \leq n$, is clear, and also (due to orthogonality) these vectors do not coincide. We prove by induction that x_i, $1 \leq i \leq n$, which by definition are orthonormal, are eigenvectors of A. Moreover, the corresponding eigenvalue λ_i is the value of the ith optimal objective function, $1 \leq i \leq n$. All of these have already been proven for the case $i = 1$. We next consider the ith optimization problem, while invoking the induction hypothesis that the result holds for all solutions up to and through $i - 1$.

Solving this problem, again by the Lagrange multipliers technique, we get that

$$L(x, \lambda_i, \mu_{ij}; 1 \leq j \leq i - 1) = x^T A x - \lambda_i(x^T I x - 1) - \sum_{j=1}^{i-1} \mu_{ij} x_j^T x.$$

Hence, for the optimal values for x and for the Lagrange multipliers,

$$\nabla_x L(x, \lambda_i, \mu_{ij}; 1 \leq j \leq i - 1) = 2Ax_i - 2\lambda_i x_i - \sum_{j=1}^{i-1} \mu_{ij} x_j = \underline{0} \in R^n.$$

$$(4.4)$$

Multiplying this from the left by x_j^T, $1 \leq j \leq i - 1$, we get that

$$2x_j^T A x_i - 0 - \mu_{ij} = 0, \quad 1 \leq j \leq i - 1.$$

Invoking the induction hypothesis for $1 \leq j \leq i - 1$, we get that

$$2\lambda_j x_j^T x_i - \mu_{ij} = 0, \quad 1 \leq j \leq i - 1.$$

Since by definition x_j and x_i are orthogonal, we conclude that $\mu_{ij} = 0$, $1 \leq j \leq i - 1$. Thus, (4.4) reduces to

$$A x_i - \lambda_i x_i = \underline{0} \in R^n,$$

as expected. The fact that λ_i is the value of objective function follows. Finally, note that as the feasible set shrinks when i increases, the optimal objective values are non-increasing. In particular, $\lambda_1 \geq \lambda_2 \geq \cdots \geq \lambda_n$. Finally, note that ties in the values of the eigenvalues are possible. □

Note that for the above objective functions, one can replace the "max" requirement by "min". All derivations are the same. The only difference is that now the eigenvalues are derived on an ascending order (and not in a descending order as above).

Sometimes, the square matrix $A \in R^{n \times n}$, in fact, equals $X^T X$ for some matrix $X \in R^{m \times n}$ and some m. Note that in this case, A is non-negative, and hence all its eigenvalues are non-negative (as can be deduced from the objective function $x^T A x$ defined in the above proof). In this case, finding the eigenvectors of A in the non-decreasing order of the value of their eigenvalues is known as the *principle component analysis* (PCA) of the matrix X. For more on PCA see, e.g., Section 10.6 in [14].

Example 7 (Simple linear regression model). Let the n pairs of points (x_i, Y_i), $1 \leq i \leq n$, be related as follows:

$$Y_i = a x_i + b + \epsilon_i, \quad 1 \leq i \leq n,$$

for some parameters a and b and some so-called *random errors* ϵ_i, $1 \leq i \leq n$, where

- The errors ϵ_i satisfy $E(\epsilon_i) = 0$, $1 \leq i \leq n$, and $E(\epsilon_i \epsilon_j) = 0$, $1 \leq i \neq j \leq n$. In particular, the errors are uncorrelated.
- $Var(\epsilon_i) = \sigma^2$, $1 \leq i \leq n$, for some σ^2 (these are known as *homoscedastic* errors).

The goal is to estimate a and b with some linear factions of the random variables Y_i, $1 \leq i \leq n$, namely functions of the type $\Sigma_{i=1}^n w_i Y_i$ from some constants w_i, $1 \leq i \leq n$. It is assumed that the value of σ^2 is given. Note that each of the desired coefficients w_i, $1 \leq i \leq n$, can be a function of x_j, $1 \leq j \leq n$, but not of the n random variables Y_i, $1 \leq i \leq n$. As each of the terms $ax_i + b$, $1 \leq i \leq n$, is not random, half the variance of the estimator is

$$\frac{\sigma^2}{2} \sum_{i=1}^n w_i^2. \tag{4.5}$$

Of course, one wishes to minimize this variance. Are there any natural constraints on w_i, $1 \leq i \leq n$, as otherwise making them all equal to zero would have been optimal (and the estimator meaningless)? Indeed there is, and it is called the unbiasedness requirement. Specifically, consider an estimator for b. It is an *unbiased estimator* (UBE) if $E(\Sigma_{i=1}^n w_i Y_i) = b$. To this end, it is needed that for any pair a and b,

$$b = E\left(\sum_{i=1}^n w_i Y_i \right) = \sum_{i=1}^n w_i E(Y_i) = \sum_{i=1}^n w_i(ax_i + b) = a \sum_{i=1}^n w_i x_i + b \sum_{i=1}^n w_i.$$

Thus, it is required that

$$\sum_{i=1}^n w_i = 1 \quad \text{and} \quad \sum_{i=1}^n w_i x_i = 0.$$

In other words, we face the problem of minimizing the separable quadratic function (4.5) under two linear constraints. Note that the factor σ^2 in the objective function can be ignored. The resulting estimator is known as the *best linear unbiased estimator* (BLUE). We next derive this estimator for b.

The Lagrangian function is

$$L(w_1, \ldots, w_n; \lambda_1, \lambda_2) = \frac{1}{2} \sum_{i=1}^n w_i^2 - \lambda_1 \left(\sum_{i=1}^n w_i - 1 \right) - \lambda_2 \sum_{i=1}^n w_i x_i.$$

Taking derivatives with respect to w_i, $1 \leq i \leq n$, and setting them to zero yields

$$w_i - \lambda_1 - \lambda_2 x_i = 0, \quad 1 \leq i \leq n. \tag{4.6}$$

Summing up these n equations, and remembering that $\Sigma_{i=1}^n w_i = 1$, yields

$$1 = n\lambda_1 + \lambda_2 \sum_{i=1}^n x_i.$$

Also, multiplying both sides of equation i in (4.6) by x_i, $1 \leq i \leq n$, and summing up the results, we get (since $\Sigma_{i=1}^n w_i x_i = 0$)

$$0 = \lambda_1 \sum_{i=1}^n x_i + \lambda_2 \sum_{i=1}^n x_i^2.$$

We get two equations for λ_1 and λ_2. It is easy to check that the solutions are

$$\lambda_1 = \frac{\sum_{i=1}^n x_i^2}{n \sum_{i=1}^n x_i^2 - (\sum_{i=1}^n x_i)^2} \quad \text{and} \quad \lambda_2 = -\frac{\sum_{i=1}^n x_i}{n \sum_{i=1}^n x_i^2 - (\sum_{i=1}^n x_i)^2}.$$

Plugging these values for the Lagrange multipliers into (4.6) implies that

$$w_j = \frac{\sum_{i=1}^n x_i^2 - x_j \sum_{i=1}^n x_i}{n \sum_{i=1}^n x_i - (\sum_{i=1}^n x_i)^2}, \quad 1 \leq j \leq n.$$

In particular, the coefficients w_j are proportional to $\overline{x^2} - x_j \overline{x}$, $1 \leq j \leq n$, where $\overline{x} = \frac{1}{n} \Sigma_{i=1}^n x_i$ and $\overline{x^2} = \frac{1}{n} \Sigma_{i=1}^n x_i^2$. Finally, to obtain the estimator for b, we need to compute $\Sigma_{j=1}^n w_j Y_j$. Some algebra implies that what we get coincides with (2.18) (where $\overline{Y}$ replaces $\overline{y}$).

The derivation for the BLUE for a is similar. In fact, the only difference is that the constraints are swapped. Specifically, the constraints are now

$$\sum_{i=1}^n w_i = 0 \quad \text{and} \quad \sum_{i=1}^n w_i x_i = 1.$$

We leave this derivation as an exercise for the reader. For completeness, we state the final answer:

$$w_j = \frac{x_j - \frac{1}{n} \sum_{i=1}^n x_i}{n \sum_{i=1}^n x_i^2 - (\sum_{i=1}^n x_i)^2}, \quad 1 \leq j \leq n.$$

In particular, the coefficients w_j are proportional to the residuals $x_j - \overline{x}$, $1 \leq j \leq n$. Again, the actual estimator for a is $\Sigma_{j=1}^n w_j Y_j$, which we claim (without proof) coincides with (2.17).

The regression line is the MLE. It is easy to see that the likelihood function here is

$$L(Y_1, \ldots, Y_n; a, b, \sigma^2) = \prod_{i=1}^n \frac{1}{\sqrt{2\pi}\sigma} e^{-\frac{(Y_i - b - a x_i)^2}{2\sigma^2}}$$

$$= \frac{1}{(2\pi)^{n/2}\sigma^n} e^{-\sum_{i=1}^n \frac{(Y_i - a - b x_i)^2}{2\sigma^2}},$$

which are indeed maximized by the regression line coefficients. Note that these estimators are free of σ^2, so whether or not this parameter is given is not an issue.

It is interesting to note that the three approaches, one in which the focus in on least-squares (and no probability model is assumed) and two others, BLUE and MLE, that are based on probability models that assume random errors, lead to the same line as the resulting solution. This is known as the Gauss–Markov theorem.

For the sake of completion, we derive the MLE for σ^2. The log of the likelihood function equals

$$-\frac{n}{2}\log(2\pi) - \frac{n}{2}\log\sigma^2 - \sum_{i=1}^{n}\frac{(Y_i - b - ax_i)^2}{2\sigma^2}.$$

Taking derivatives (separately) with respect to a, b, and σ^2, and equating them to zero, leads to the above MLEs for a and b, denoted next by $\hat{a}$ and $\hat{b}$, respectively. Then, the MLE for σ^2 equals

$$\hat{\sigma}^2 = \frac{1}{n}\sum_{i=1}^{n}(Y_i - \hat{b} - \hat{a}x_i)^2.$$

Finally, if a and/or b are given in the above estimator for σ^2, their estimators are repalced by their actual value. For more on estimating parameters in the case of simple linear regression model, see [11, pp. 539–563].

Example 8 (A counterexample). In this example, the gradients of the constraints are not linearly independent, namely the optimal point is not regular[b]:

$$\min_{x_1, x_2} \quad x_1 + x_2,$$

$$\text{s.t. } (x_1 - 1)^2 + x_2^2 - 1 = 0,$$

$$(x_1 - 2)^2 + x_2^2 - 4 = 0.$$

It is easy to see that there exists a unique feasible point $x^* = (0,0)$, which is automatically optimal. $\nabla f(x^*) = (1,1)$, while $\nabla g_1(x^*) = (-2,0)$ and $\nabla g_1(x^*) = (-4,0)$. It is easy to see that $\nabla f(x^*)$ cannot be expressed as a linear combination of $\nabla g_1(x^*)$ and $\nabla g_2(x^*)$. Indeed, observe that $\nabla g_1(x^*)$ and $\nabla g_2(x^*)$ are clearly not linearly independent. This example shows that

---

[b]This example appears in [5, p. 348].

the FOCs are not necessarily obeyed by the optimal point in the case where the regularity conditions are not met at this point.

Exercise 4.1

Solve the following constrained optimization problem

$$\min_{x \in R^n} \frac{1}{2} \sum_{i=1}^{n} x_i^2,$$

$$\text{s.t. } \sum_{i=1}^{n} x_i = c.$$

From the solution, deduce that

$$\frac{1}{n} \sum_{i=1}^{n} x_i^2 \geq \left(\frac{1}{n} \sum_{i=1}^{n} x_i \right)^2,$$

with equality if and only if $x_1 = x_2 = \cdots = x_n$.

Exercise 4.2

Consider the m-constraint optimization problem

$$\min_{x} f(x),$$

$$\text{s.t. } g_i(x) = 0, \quad 1 \leq i \leq m,$$

and the one-constraint optimization problem

$$\min_{x} f(x),$$

$$\text{s.t. } \sum_{i=1}^{m} g_i^2(x) = 0.$$

1. Explain why the two problems are equivalent.
2. Explain why the Lagrange multiplier technique is not going to work for the latter problem.

Exercise 4.3

Assume that a symmetric matrix $A \in R^{n \times n}$ has n orthonormal eigenvectors $w_1, \ldots, w_n$, arranged by columns in a matrix W, with corresponding eigenvalues $\lambda_1 \leq \cdots \leq \lambda_n$. Let D be a diagonal matrix, with all the above eigenvalues in its main diagonal. Show that[c]:

(a) $W^T W = I$.
(b) $\max_{\|x\|=1} x^T A x = \max_{\|x\|=1} x^T D x = \lambda_n$.
(c) State the corresponding result when "min" is replaced by "max".
(d) Assume now that $\lambda_1 > 0$. Show that

$$\frac{1}{2} = \min_{0 \leq \alpha \leq 1} \max \frac{\frac{1}{\alpha\lambda_1+(1-\alpha)\lambda_n}}{\alpha\frac{1}{\lambda_1} + (1-\alpha)\frac{1}{\lambda_n}}.$$

Note that one maximizes the reciprocal of the function given here. Conclude that the minimum value equals $\frac{4\lambda_1\lambda_n}{(\lambda_1+\lambda_n)^2}$, which is the square of the ratio of the geometric mean to the algebraic mean between λ_1 and λ_n. It is well-known that this ratio is smaller than or equal to 1.

4.2.2 A proof of Theorem 4.1

For the single variable $t \in R$, define n differentiable (and hence continuous) functions of t, $x(t) = (x_1(t), x_2(t), \ldots, x_n(t)) \in R^n$. Assume that $x(0) = x^*$ and that for some open neighborhood around $t = 0$, all points $x(t) \in R^n$ are feasible points, namely $g_i(x(t)) = 0$, $1 \leq i \leq m$. For obvious reasons, these n functions define what is called a *feasible path*.[d] Clearly, a necessary condition for the optimality of x^* is that $f(x(t))$, as a function of t, receives its minimum value at the point $t = 0$. For that, one needs $\frac{df(x(t))}{dt}|_{t=0} = 0$ for any feasible path. But, by (2.4),

$$\frac{df(x(t))}{dt} = \sum_{i=1}^n \frac{df(x(t))}{dx_i} \frac{dx_i(t)}{dt} = \nabla f(x(t))^T \dot{x}(t),$$

---

[c]In contrast to Theorem 4.2 above, existence is assumed here.
[d]For a given function $h(x) : R^n \to R$ and some domain for $t \in R$, all points $x(t) \in R^n$ such that $h(x(t)) = c$ for some constant c are referred to an *indifference curve* or *contour level*. See Figure 4.1.

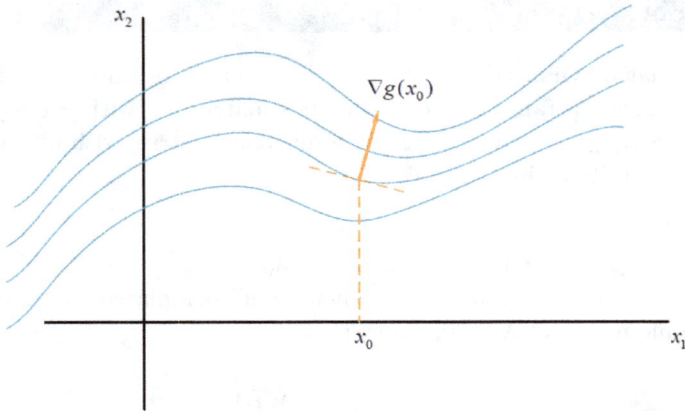

Figure 4.1. Indifference curves and a gradient.

where $\dot{x}(t) \in R^n$ denotes the vector of the n derivatives of the n functions $x_i(t)$ with respect to t, $1 \leq i \leq n$. Insert $t = 0$, then, in order for this value to be optimal for t, it is required that

$$\nabla f(x^*)^T \dot{x}(0) = 0.$$

In words, we need the gradient of the function at x^* be orthogonal to all directional derivatives generated by all feasible paths. Note that these direction derivatives forms a linear subspace: If for two feasible paths $x(t)$ and $y(t)$, $\nabla f(x^*)^T \dot{x}(0) = 0$ and $\nabla f(x^*)^T \dot{y}(0) = 0$, then $\nabla f(x^*)^T (\alpha \dot{x}(0) + \beta \dot{y}(0)) = 0$ for any pair of α and β.

By a linear algebra argument already used, we claim that this equality condition implies that $\nabla f(x^*)^T$ lies in the linear subspace spanned by the set of vectors that are orthogonal to all such directional derivatives, and hence it can be expressed as a linear combination of a set of vectors spanning it, namely form a basis for it. You may recall the definition of the tangent plane: the linear subspace of vectors $d \in R^n$ such that $\nabla g(x)d = \underline{0} \in R^m$. We claim that under the stated regularity condition, the set of directional vectors derived from all possible paths coincide with the tangent plane. In what follows, we validate this claim.

Firstly, if $x(t)$ is a feasible path, then by definition, $g_i(x(t)) = 0$, $1 \leq i \leq m$, for some neighborhood of $t = 0$. Hence,

$$0 = \frac{dg_i(x(t))}{dt} = \sum_{i=1}^{n} \frac{dg_i(x(t))}{dx_i} \frac{dx_i(t)}{dt} = \nabla g_i(x(t))^T \dot{x}(t), \quad 1 \leq i \leq m.$$

In particular,

$$0 = \nabla g_i(x(0))^T \dot{x}(0) = \nabla g_i(x^*)^T \dot{x}(0), \quad 1 \le i \le m,$$

establishing the fact that $\dot{x}(0)$ is in the tangent plane for any feasible path.

Conversely, assume that $d \in R^n$ is in the tangent plane, namely $\nabla g_i(x^*)^T d = 0$, $1 \le i \le m$. Then construct a feasible path $x(t)$ with $\dot{x}(0) = d$. To this end, consider the following m constraints with $m + 1$ variables $t \in R$ and $u \in R^m$:

$$\underline{g}(x^* + td + \nabla \underline{g}(x^*)^T u) = \underline{0} \in R^m.$$

A feasible solution is of course $t = 0$ and $u = \underline{0} \in R^m$. If we look at the neighborhood of $t = 0$ and see when t varies, we can see how u varies accordingly, we in fact get the m continuous functions $u(t) : R \to R^m$ around $t = 0$ that satisfy the identity

$$\underline{g}(t, u) = \underline{g}(x^* + td + \nabla \underline{g}(x^*)u(t)) = \underline{0} \in R^m.$$

The implicit function theorem (see Appendix in Section 4.6 below) implies that this task is possible when the required regularity conditions hold here.[e] Using the chain rule, we can argue that

$$\nabla_u \underline{g}(t, u)|_{t=0} = \nabla \underline{g}(x^*)^T \nabla \underline{g}(x^*) \in R^{m \times m},$$

which is invertible due to the fact that the m columns of $\nabla g(x^*)$ are linearly independent. The next thing we argue is that $\dot{u}(0) = \underline{0} \in R^m$. Indeed, again by the chain rule,

$$R^m \ni \underline{0} = \dot{\underline{g}}(x(t))|_{t=0} = \nabla g(x^*)^T (d + \nabla g(x^*)\dot{u}(0))$$

$$= \underline{0} + \nabla g(x^*)^T \nabla g(x^*)\dot{u}(0).$$

The fact that $\nabla \underline{g}(x^*)^T \nabla g(x^*) \in R^{m \times m}$ is regular completes the argument. Finally, by construction, $x(t) = x^* + td + \nabla g(x^*)u(t)$ is a feasible set. Clearly, $\dot{x}(0) = \nabla \underline{g}(x^*)\dot{\underline{g}}(0) = \underline{0} \in R^n$. This completes the proof.

---

[e]We have $m + 1$ variables: t and $u_1, \ldots, u_m$, and the m equality conditions lead to the idea that m of these variables (namely $u_1, \ldots, u_n$) are implicitly m functions of one of these variables (in this case, the variable t).

4.3 Second-Order Conditions (SOCs)

When we consider the unconstrained optimization of $f(x) : R^n \to R$, once a point x was observed to be stationary, namely $\nabla f(x) = \underline{0}$, the next question is whether or not the Hessian $\nabla^2 f(x)$ is positive, negative, or none of the above, which should lead to the conclusion that x is a local minimum, a local maximum, or undecided, respectively. What is the corresponding condition in the case of constrained optimization? It is tempting to jump to the conclusion that a condition similar to the one used for the unconstrained optimization case holds here too, but it will be less severe in the sense that such a positive or negative requirement is required only for feasible directions, namely one needs to check the sign of $d^T \nabla^2 f(x) d$ for all $d \neq \underline{0}$ with $\nabla g_i(x) d = 0$, $1 \leq i \leq m$. But as it turns out, $d^T \nabla^2 f(x) d$ needs to be replaced with

$$d^T \left(\nabla^2 f(x) - \sum_{i=1}^{m} \lambda_i \nabla^2 g_i(x) \right) d,$$

where $\lambda = (\lambda_1, \ldots, \lambda_m)$ is the vector of Lagrange multipliers associated with x.

Theorem

Theorem 4.3. *Let x^* and λ be a pair consisting of a feasible solution and a vector of Lagrange multipliers that meet the regularity condition and FOCs stated in Theorem 4.1. Then, a necessary condition for local minimum at x^* is that*

$$d^T \left(\nabla^2 f(x^*) - \sum_{i=1}^{m} \lambda_i \nabla^2 g_i(x^*) \right) d \geq 0, \qquad (4.7)$$

for all feasible directions $d \neq \underline{0}$ at x^*. Moreover, for sufficiency, one needs to replace "non-negative" here with "positive".

Proof. Recall the definition of $f(x(t))$ over a feasible path $x(t)$ that crosses x^* (where $t = 0$). A necessary condition for x^* to be a local minimum is that the second derivative of $f(x(t))$ (with respect to t) is non-negative when $t = 0$; however, it needs to be positive to be a sufficient condition for optimality. By the chain rule,

$$\frac{d^2 f(x(t))}{dt^2} = \dot{x}(t)^T \nabla^2 f(x(t)) \dot{x}(t) + \nabla f(x(t))^T \ddot{x}(t).$$

Similarly,

$$\frac{d^2 \sum_{i=1}^n \lambda_i g_i(x(t))}{dt^2}$$

$$= \dot{x}(t)^T \left[\sum_{i=1}^n \lambda_i \nabla^2 g_i(x(t)) \right] \dot{x}(t) + \left[\sum_{i=1}^n \lambda_i \nabla g_i(x(t))^T \right] \ddot{x}(t).$$

Note that the last of the above expressions is zero for any t in the neighborhood of $t = 0$ due to the feasibility of the path (making $g_i(x(t)) = 0$ an identity, $1 \leq i \leq m$, along such a path). Combining the above two expressions we get, after some quick rearrangement, that

$$\frac{d^2 f(x(t))}{dt^2} = \dot{x}(t)^T \left[\nabla^2 f(x(t)) - \sum_{i=1}^n \lambda_i \nabla^2 g_i(x(t)) \right] \dot{x}(t)$$

$$+ \left[\nabla f(x(t) - \sum_{i=1}^n \lambda_i \nabla g_i(x(t))) \right]^T \ddot{x}(t).$$

Now insert $t = 0$. The second term vanishes since x^* and λ are assumed to meet the FOCs, while the first term equals

$$\dot{x}(0)^T \left[\nabla^2 f(x^*) - \sum_{i=1}^n \lambda_i \nabla^2 g_i(x^*) \right] \dot{x}(0).$$

In particular, we get a matrix with non-negativity (respectively, positivity) along the tangent plane (and $\dot{x}(0)$ is in the tangent plane) which is required for a necessary (respectively, sufficient) condition for local minimum at x^*.
□

Remark. Note that $\nabla^2 f(x) - \sum_{i=1}^n \lambda_i \nabla^2 g_i(x)$ is in fact $\nabla_{xx} L(x, \lambda)$.

Definition

Definition 4.4. A feasible point x^* that obeys the conditions stated in (4.7) is said to meet the second-order conditions (SOCs).

Example 1 (cont.). Is the point $(1/3, 2/3, 1/3)$ a local minimum, a maximum, or neither? Let us look at the second-order conditions. First, the Hessian of the objective function is the identity matrix. The zero matrix is the Hessian of the two constraints. The identity matrix is clearly a positive matrix, therefore it is positive for the feasible directions stemming from the stationary point $(1/3, 2/3, 1/3)$. Hence, this point is a local minimum.

Example 9. Consider the following constrained optimization:

$$\text{optimize}_{x_1, x_2. x_3} \ x_1 x_2 x_3,$$

$$\text{s.t. } x_1 x_2 + x_1 x_3 + x_2 x_3 = 1.$$

The gradient of the objective function equals $(x_2 x_3, x_1 x_3, x_1 x_2)$. The gradient of the single constraint is $(x_2 + x_3, x_1 + x_3, x_1 + x_2)$. Thus, we look for (x_1, x_2, x_3) and a multiplier λ with

$$x_2 x_3 - \lambda(x_2 + x_3) = 0,$$

$$x_1 x_3 - \lambda(x_1 + x_3) = 0,$$

$$x_1 x_2 - \lambda(x_1 + x_2) = 0,$$

$$x_1 x_2 + x_2 x_3 + x_1 x_3 = 1.$$

The unique solution is $x_1 = x_2 = x_3 = \frac{\sqrt{3}}{3}$ with $\lambda = \frac{\sqrt{3}}{6}$.

Is this a local minimum, a local maximum, or neither? Let us look at the second-order conditions. First, the Hessian of the objective function at the stationary point is

$$\nabla_x^2 f(x)\big|_{x_1=x_2=x_3=\frac{\sqrt{3}}{3}} = \begin{pmatrix} 0 & \frac{\sqrt{3}}{3} & \frac{\sqrt{3}}{3} \\ \frac{\sqrt{3}}{3} & 0 & \frac{\sqrt{3}}{3} \\ \frac{\sqrt{3}}{3} & \frac{\sqrt{3}}{3} & 0 \end{pmatrix}.$$

The Hessian of the single constraint at the stationary point (or at any other point) equals

$$\begin{pmatrix} 0 & 1 & 1 \\ 1 & 0 & 1 \\ 1 & 1 & 0 \end{pmatrix},$$

which up to a multiplicative constant coincides with the Hessian of the objective function at the stationary point. Finally, $\lambda > 0$.

Thus, the matrix we need to consider while checking for the SOCs, is

$$\frac{\sqrt{3}}{6} \begin{pmatrix} 0 & 1 & 1 \\ 1 & 0 & 1 \\ 1 & 1 & 0 \end{pmatrix}.$$

We need to check the above matrix, denote it by Q, is positive, negative, or neither, for all directions that are orthogonal to $(1, 1, 1)$. In other words, we need to check the sign of $d^T Q d$ for all non-zero d that satisfy $d_1 + d_2 + d_3 = 0$. Thus, up to a positive constant, $d^T Q d = d_1(d_2 + d_3) + d_2(d_1 + d_3) + d_3(d_1 + d_2) = -d_1^2 - d_2^2 - d_3^2$, which is negative. Thus, the stationary point is a local maximum.

The above two examples are cases in which one minimizes a convex function under affine equality constraints. Since in this case $\nabla^2 g_i(x)$ is the $n \times n$ zero matrix, $1 \leq i \leq m$, the SOCs are automatically met. In other words, in this case one needs to consider only the FOCs.

Theorem

Theorem 4.4. *Assume that $f(x)$ is convex and that $g_i(x)$ is affine, $1 \leq i \leq m$. Then, if x^* meets the FOCs for (4.1), then x^* is a globally optimal solution.*

Proof. Suppose that $x^* \in R^n$ and $\lambda^* \in R^m$ meet the FOCs. Define the function $h(x) : R^n \to R$ as

$$h(x) = f(x) - (\lambda^*)^T g(x).$$

Clearly, $h(x) = f(x)$ for any feasible x. Also, it follows that for any x, $\nabla_x^2 h(x) = \nabla_x^2 f(x)$, implying that $h(x)$ is a convex function. As stated in Theorem 2.8, for convex functions, the FOCs are sufficient for global optimization. Indeed,

$$\nabla_x h(x) = \nabla f(x) - (\lambda^*)^T \nabla g(x),$$

which equals $\underline{0}$ at x^*. $\qquad\square$

As we will show in the next chapter (see Theorem 5.3), the key property in the above theorem is that the feasible set is convex. In presence of equality constraints this property is guaranteed if and only if the functions defining the constraints are affine.

Example 10 (The steepest descent direction (revisited)). $-\nabla f(x)$ is uniquely the steepest descent direction of the function $f(x) : R^n \to R$ at the point x.[f]

---

[f]This is similar to Theorem 2.5, though a different proof is given.

Proof. Denote by $d^* \in R^n$ the normalization of $\nabla f(x)$, namely $d^* = \nabla f(x)/||\nabla f(x)||$. We need to show that

$$-d^* = \arg \min_{d \in R^n} \{\nabla f(x)^T d, \quad \text{s.t.} \quad ||d||^2 = 1\}.$$

The Lagrangian function is

$$L(d, \lambda) = \nabla f(x)d - \lambda(||d||^2 - 1).$$

The FOCs are

$$\nabla f(x) - 2\lambda d = 0,$$

namely requiring the optimal solution d to be proportional to $\nabla f(x)$, regardless of whether we seek a minimization or maximization point. Specifically, up to a positive multiplicative constant, the only two possible solutions are

$$(d, \lambda) = (d^*, ||\nabla f(x)||/2) \quad \text{and} \quad (d, \lambda) = (-d^*, -||\nabla f(x)||/2).$$

The SOCs are

$$\nabla_d^2 L(d, \lambda) = -2\lambda I.$$

Thus, the sufficient SOCs for maximization (respectively, minimization), hold at the first (respectively, maximization) solution.   □

Exercise 4.4

This exercise is similar to Exercise 3.3.2 but has an additional condition: The value of service completion is server dependent. Specifically, the value of service granted by server i is denoted by α_i, $1 \leq i \leq n$.[g] Answer the following questions:

1. What is the objective function now? What are the constraints?
2. Show that the objective function is concave function with respect to any of its variables (while all others are fixed).
3. Add Lagrange multipliers and state the KTT conditions for optimality.

---

[g]This problem appears in [1].

4. Order the servers in a descending order of α_i. Show that in the optimal solution server i is open, then so is server $i - 1$. In particular, for some i^*, only the first i^s servers are open. Note that $i^* = n$ is possible.
5. How does α_{i^*+1} compare with the Lagrange multiplier of the equality constraint?
6. Assume that i^* is known. What is the optimal split of the arrival stream and what is the corresponding value of the objective function?
7. State a procedure for finding i^*.

4.4   Sensitivity Analysis

In examining constrained optimization under equality constraints, we looked for feasible points that meet FOCs and SOCs. In particular, we were interested in the existence of numbers, called the Lagrange multipliers, that coupled with the solution, meet a number of conditions. So far, we were not interested in their actual values themselves, and moreover, we saw that they possess only a theoretical value. We now show that much can be learned from these values.

Specifically, suppose that for some real number c, the constraint $g_i(x) = 0$ is changed to $g_i(x) = c$ for some i, $1 \leq i \leq m$, while the rest of the optimization problem remains untouched. This change of course leads to changes in the optimal values for the variables and in the objective function itself. In particular, if the above imposed regularity constraints are met when $c = 0$, we expect these values to change continuously with c. Thus, define $h(c)$ as the optimal objective value as a function of c. Note that this is a differentiable function with $h(0) = f(x^*)$. Our main claim here is that

$$\frac{d\,h(c)}{d\,c}\Big|_{c=0} = \lambda_i,$$

where λ_i is the Lagrange multiplier of the ith constraint of the original problem. This claim deals with a typical sensitivity analysis issue: It shows how much a function changes due to a small change (sometimes referred to as a perturbation) which takes place in one of its input parameters. It shows how much, infinitesimally, the objective function value changes when considered as a function of a right-hand side entry. What we are claiming here is that this change equals the product between the corresponding Lagrange multiplier and the (infinitesimal) magnitude of the change in the right-hand side value.

Theorem

Theorem 4.5. *Consider the following equality constrained optimization problem*

$$\min_{x \in R^n} f(x),$$

$$s.t. \quad g(x) = c \in R^m,$$

for some $m < n$. Denote by $x(c)$ and $\lambda(c)$ a solution and the corresponding Lagrange multipliers who meet the FOCs under c. Assume that both $x(c)$ and $\lambda(c)$ exist for some neighborhood of $c = \underline{0}$ and that both are differentiable with respect to c. Also, assume that the regularity assumptions imposed above hold for $c = \underline{0}$. Then,

$$\nabla_c f(x(c))|_{c=\underline{0}} = \lambda(\underline{0}). \tag{4.8}$$

Proof. Suppose that the pair of differentiable functions $(x(c), \lambda(c))$ solve the following $(m + n) \times (n + m)$ non-linear equations:

$$\nabla f(x) - \nabla \underline{g}(x)^T \lambda = \underline{0} \in R^n \quad \text{and} \quad \underline{g}(x) = c \in R^m. \tag{4.9}$$

In other words, they meet the FOCs when c is the right-hand side vector. Let $x^* \in R^n$ and $\lambda^* \in R^m$ be a solution where $c = \underline{0}$.

By the chain rule, we get that

$$\nabla_c f(x(c))|_{c=\underline{0}} = \nabla_c x(\underline{0})^T \nabla f(x^*),$$

and

$$\nabla_c \underline{g}(x(c))|_{c=\underline{0}} = \nabla_c x(\underline{0})^T \nabla \underline{g}(x^*)^T.$$

Note that $\nabla_c x(c)$ and $\nabla_c \underline{g}(x(c))$ are the Jacobians of $x(c) : R^m \to R^n$ and $\underline{g}(x(c)) : R^m \to R^m$, respectively. It is clear that since $\underline{g}(x(c)) = c$, it follows that $\nabla_c \underline{g}(x(c))|_{c=\underline{0}} = I \in R^{m \times m}$, which can be placed on the left-hand side of the latter equality. Multiplying the latter identity by λ^* from the right and subtracting this sum from the former equality, we get that

$$\nabla_c f(x(c))|_{c=\underline{0}} - \lambda^* = \nabla_c x(\underline{0})^T \nabla f(x^*) - \nabla_c c(\underline{0})^T \nabla g(x^*)^T \lambda^*$$
$$= \nabla_c x(\underline{0})^T (\nabla f(x^*) - \nabla g(x^*)^T \lambda^*).$$

Finally, (4.9) implies that the right-hand side is in $\underline{0} \in R^m$. This completes the proof. ☐

Theorem 4.5 is usually referred to in the literature as the *envelope theorem*.

Remark. The function $\lambda(c)$ can be looked at as an implicit function of $x(c)$ around the point (x^*, λ^*). See Section 4.6 for details.

4.4.1 *Queueing Examples*

Example 1. You may recall the problem of optimizing the splitting of the total service capacity among servers introduced and solved in Section 3.3.1. It was also revisited as Example 4 in Section 4.2.1. In particular, it was shown there that if a total service capacity of μ is split in an optimal way among n servers so as to minimize the mean sum of the queue lengths, then the optimal value equals

$$\frac{(\sum_{i=1}^n \sqrt{\lambda_i})^2}{\mu - \sum_{i=1}^n \lambda_i}.$$

See (3.7). The derivative of this function with respect to μ equals

$$-\frac{(\sum_{i=1}^n \sqrt{\lambda_i})^2}{(\mu - \sum_{i=1}^n \lambda_i)^2},$$

which indeed coincides with the Lagrange multiplier that corresponds to the equality constraint $\sum_{i=1}^n x_i = \mu$ as it appeared in (4.3). In this example, it is also possible to compute the derivatives of the optimal variables with respect to the right-hand side. Specifically, by (3.6), we get that at the optimal solution,

$$\frac{d\,x_i}{d\,\mu} = \frac{\sqrt{\lambda_i}}{\sum_{j=1}^n \sqrt{\lambda_j}}, \quad 1 \le i \le n.$$

Example 2. You may recall the problem of optimizing the splitting of the total arrival stream of size λ among various servers that was introduced and solved in Section 3.3.2. It was shown there that the optimal objective value equals

$$\frac{(\sum_{j=1}^{i} \sqrt{\mu_j})^2}{\sum_{j=1}^{i} \mu_j - \lambda} - i,$$

where i is the index of the slowest open server. See (3.10). Its derivative with respect to λ clearly equals

$$\frac{(\sum_{j=1}^{i} \sqrt{\mu_j})^2}{(\sum_{j=1}^{i} \mu_j - \lambda)^2},$$

that coincides with the Lagrange multiplier of the equality constraint as it appeared in (3.9). From (3.8), we learn that at the optimal solution

$$\frac{d\,x_j}{d\,\lambda} = \frac{\sqrt{\mu_j}}{\sum_{k=1}^{i} \sqrt{\mu_k}}, \quad 1 \leq j \leq i.$$

Clearly,

$$\frac{d\,x_j}{d\,\lambda} = 0, \quad i < j \leq n.$$

Somewhat surprisingly, these derivatives, as long as i stays fixed, are not functions of λ. In other words, all individual optimal arrival rates increase linearly with λ, albeit each at its own pace. As it should be, the sum of these derivatives equals 1.

Remark. The Lagrange multipliers are also called *shadow prices*. The reason behind this can be seen from (4.8). Specifically, for some i, $1 \leq i \leq m$, consider the constraint $g_i(x) = 0$. Interpret this constraint as limiting the amount of some resource that can be used as part of some production process. Suppose one wishes to relax it to $g_i(x) = \epsilon$ or $g_i(x) = -\epsilon$. One may wish to do so in order to improve the value of optimal objective function. The value of the Lagrange multiplier says what (infinitesimally) the gain (or loss) in the objective function per unit of change (or perturbation) in the resource value is. This is exactly how much one is willing to pay for such a change per unit of change or, equivalently, to be paid for per unit in reducing what one currently possesses in this resource.

Remark. There is a school of thought that in practice, especially when money matters are involved, there are actually no constraints. Why does some commodity need to be equal to, say, 10.4 and not 10.5? At some price we can always have 10.5 if we wish. Thus, the constraint is somewhat artificial, while what really matters is the shadow price, which tells what the effect is on the value of the optimal objective function due to this change of 0.1.

Exercise 4.5

Consider the following optimization problem:

$$\min_{x_1, x_2, x_3} x_1 + x_2 + x_3,$$

$$\text{s.t.} \quad x_1^2 + x_2 = 3,$$

$$x_1 + 3x_2 + 2x_3 = 7.$$

1. Show that this problem is equivalent to the unconstrained optimization problem $\min_{x_1}(x_1^2 + x_1 + 4)/2$. Solve this problem and deduce the optimal solution to the original problem.
2. Write the Lagrangian function and the KKT conditions for the original problem.
3. Using the fact that the optimal solution is known, find the corresponding Lagrange multipliers.
4. Suppose that the 3 and the 7 in the right-hand side of the above are replaced with 3.01 and 6.98, respectively. Give an approximation for the new value optimal objective function.

4.5 Duality in Non-Linear Programming

4.5.1 *The dual problem and the dual theorem*

For simplicity, assume a constrained optimization with only equality constraints:

$$\min_{x \in R^n} f(x), \tag{4.10}$$

$$\text{s.t.} \quad g(x) = \underline{0} \in R^m.$$

In the context of duality analysis, the problem is referred to as the *primal* problem. We assume $m < n$ constraints. The FOCs we have defined convert the optimization problem into a set of constraints or, equivalently, (not necessarily linear) equations. One of the drawbacks in conversion is the introduction of another set of m unknowns, $\lambda \in R^m$. The primary focus is of course on the original n variables. An alternative approach, called here *duality* approach, is to deal first with the new variables, called the *dual variables*.

To this end, define the function $\phi(\lambda) : R^m \to R$, called the *dual function*:

$$\phi(\lambda) = \min_{x \in R^n} L(x, \lambda) = \min_{x \in R^n} \{f(x) - \lambda^T \underline{g}(x)\}.$$

Note that this function is defined via an unconstrained optimization problem.

> **Lemma**
>
> **Lemma 4.1.** *The dual function is concave.*

Proof. For α with $0 \le \alpha \le 1$,

$$\phi(\alpha\lambda_1 + (1 - \alpha)\lambda_2)$$
$$= \min_{x \in R^n} \{f(x) - (\alpha\lambda_1 + (1 - \alpha)\lambda_2)^T \underline{g}(x)\}$$
$$= \min_{x \in R^n} \{\alpha(f(x) - \lambda_1^T \underline{g}(x)) + (1 - \alpha)(f(x) - \lambda_2^T \underline{g}(x))\}$$
$$\ge \alpha \min_{x \in R^n} \{f(x) - \lambda_1^T \underline{g}(x)\} + (1 - \alpha) \min_{x \in R^n} \{f(x) - \lambda_2^T \underline{g}(x)\}$$
$$= \alpha\phi(\lambda_1) + (1 - \alpha)\phi(\lambda_2).$$

$\square$

The *dual optimization problem* is

$$\max_{\lambda \in R^m} \phi(\lambda).$$

The following theorem is known as the weak duality theorem.

> **Theorem**
>
> **Theorem 4.6.** *For any $\lambda \in R^m$ and for any feasible $x \in R^n$, $\phi(\lambda) \le f(x)$.*

Proof. Denote by x^* the globally optimal solution to the primal problem. For any $\lambda \in R^m$ and for any primal feasible $x \in R^n$,

$$\phi(\lambda) = \min_{x \in R^n} \{f(x) - \lambda^T \underline{g}(x)\} \le f(x^*) - \lambda^T \underline{g}(x^*) = f(x^*) \le f(x). \quad \square$$

Remark. The above duality theorem appears in its "full" version in the sense that all constraints are put in the objective function. It is possible to show that it also holds in its partial version. For example, it is possible to partition the set of constraints into two classes: those with indices from 1 up to k, and those with indices $k+1$ up to m. Then, for $\lambda \in R^k$, the dual function is

$$\phi(\lambda) = \min_{x \in R^n, g_i(x) = 0, k+1 \leq i \leq m} \left\{ f(x) - \sum_{i=1}^{k} \lambda_i g_i(x) \right\},$$

and, as before, the dual optimization problem is

$$\max_{\lambda \in R^k} \phi(\lambda).$$

In fact, the original version of the dual theorem is a special case of this version where $k = m$.

Set $\lambda^* = \arg\max_{\lambda \in R^m} \phi(\lambda)$. It is immediate by the weak duality theorem that $\phi(\lambda^*) \leq f(x^*)$. In the case where the inequality is strict, we say that a duality gap exists. In this case, the solution of the dual problem leads only to a lower bound on the value of the optimal primal objective function. Of course, much interest exists in the case where there is no duality gap. In particular, if one finds a feasible primal solution x and a dual solution λ with $f(x) = \phi(\lambda)$, then both solutions are optimal, x for the primal and λ for the dual. Moreover, $f(x) = \phi(\lambda)$ is the optimal value for both. Also, if one of the optimization problem is unbounded, then the other is infeasible.

Next, we state a case in which no duality gap exists.

Theorem

Theorem 4.7. *Suppose that x^* is feasible for* (4.10) *and, coupled with λ^*, it meets the FOCs. Suppose further that*

$$x^* = \arg\max_{x \in R^n} L(x, \lambda^*), \qquad (4.11)$$

then, x^* is primal optimal, λ^* is dual optimal, and no duality gap exists.

Proof.

$$\min_{x \in R^n, \underline{g}(x) = \underline{0}} f(x) \le f(x^*) = L(x^*, \lambda^*) = \max_{x \in R^n} L(x, \lambda^*)$$

$$= \phi(\lambda^*) \le \max_{\lambda \in R^m} \phi(\lambda).$$

The weak duality theorem (see Theorem 4.6) points to the reverse inequality. Hence, the inequality stated above is in fact an equality.   □

Note that the extra assumption (4.11) is not that strong: It requires that x^* (which already meets the FOCs which are necessary for optimality) is a maximum point with respect to Lagrangian function (but with the specific value for the dual variables, namely λ^*). Also note that Theorem 4.7 complements Theorem 4.4 in the sense that global optimization is established in both without the need to resort to the use of SOCs.

Remark. The solution (x^*, λ^*) for the optimization problem

$$\max_{\lambda \in R^m} \min_{x \in R^n} \{f(x) - \lambda^T \underline{g}(x)\}$$

is called a *saddle point* as it is the minimum point in one direction, namely in x, and the maximum point in another direction, namely, in λ. The connotation is clear. When you sit on a saddle placed on a horse, in one direction, where the road lies, you are at the bottom, while in the orthogonal direction, where your legs are, you are on the top. See Figure 4.2.

Remark. An iterative primal algorithm would be based on the idea of guessing for the optimal solution and then checking whether it leads to the existence of Lagrange multipliers (namely if it is dual feasible). Usually, the answer is "no", and then another feasible solution needs to be derived, the corresponding multipliers are sought for, etc. The dual approach does the opposite. Specifically, one starts with some λ. Then one solves the unconstrained optimization problem $x(\lambda) = \arg\min_{x \in R^n} \{f(x) - \lambda^T \underline{h}(x)\}$ and checks whether $x(\lambda)$ is feasible, namely whether $h(x(\lambda)) = \underline{0}$. If this is the case, one is done because the optimal solution is attained. However, usually, one is not so lucky and needs to try for a better λ, say λ_1 such that $\phi(\lambda_1) > \phi(\lambda)$. How this can be done, namely how a better dual solution is found, is another matter. For that, one constructs a series of approximations $\lambda_1, \lambda_2, \ldots$ for λ^*. By the monotonicity of $\phi(\lambda_i)$, convergence is guaranteed, and the hope is that the limit is $\phi(\lambda^*)$.

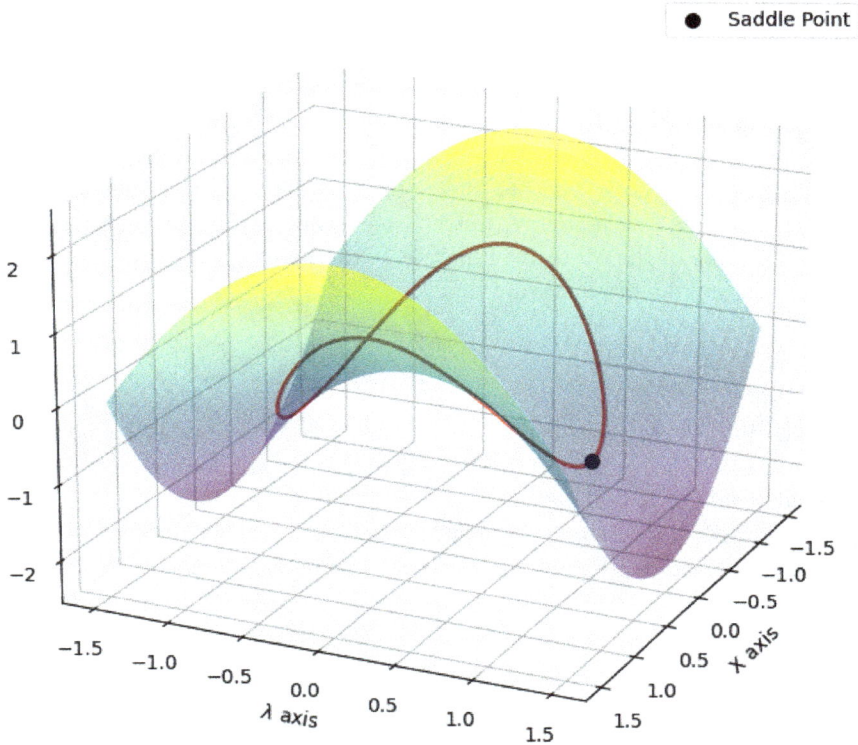

Figure 4.2. Saddle point under a constraint.

4.5.2 *Examples*

Example 1. Consider the following optimization problem under one equality constraint[h]:

$$\min_{x,y} -xy,$$

$$\text{s.t. } (x-3)^2 + y^2 - 5 = 0.$$

The Lagrangian function is

$$L(x,y,\lambda) = -xy - \lambda[(x-3)^2 + y^2 - 5].$$

---

[h]This example appears in [18, p. 450].

The FOCs are

$$-y - 2\lambda(x - 3) = 0,$$

and

$$-x - 2\lambda y = 0.$$

It is possible to check whether the point $(x, y) = (4, 2)$ is feasible and, coupled with the Lagrange multiplier $\lambda = -1$, meets the FOCs. Note that -8 is the value of the objective function at the point $(4, 2)$. For the SOCs, we need to look at the following matrix:

$$\begin{pmatrix} 0 & -1 \\ -1 & 0 \end{pmatrix} - (-1) \times \begin{pmatrix} 2 & 0 \\ 0 & 2 \end{pmatrix} = \begin{pmatrix} 2 & -1 \\ -1 & 2 \end{pmatrix}.$$

It is possible to see that this matrix is positive, for example, by noticing that its eigenvalues are 1 and 3. In particular, they are positive. Hence, we conclude that $(4, 2)$ is the optimal point.

Next, we demonstrate the dual approach to solving this problem. Specifically, for a given λ, we need to consider the dual function

$$\phi(\lambda) = \min_{x,y}\{-xy - \lambda((x - 3)^2 + y^2 - 5)\}.$$

Taking the derivatives with respect to x and y (separately) and setting them to zero yields

$$x(\lambda) = \frac{12\lambda^2}{-1 + 4\lambda^2} \quad \text{and} \quad y(\lambda) = \frac{-6\lambda}{-1 + 4\lambda^2}.$$

Plugging these two into the function $-xy - \lambda((x - 3)^2 + y^2 - 5)$ implies that

$$\phi(\lambda) = \frac{72\lambda^3}{(4\lambda^2 - 1)^2} - \lambda\left(\frac{9}{(4\lambda^2 - 1)^2} + \frac{36\lambda^2}{(4\lambda^2 - 1)^2} - 5\right).$$

The next question is $\max_\lambda \phi(\lambda)$. For this, one needs to differentiate $\phi(\lambda)$ with respect to λ. It is possible to argue that the value $\lambda = -1$ makes the derivative zero. Finally, $\phi(-1) = -8$, $x(-1) = 4$, and $y(-1) = 2$, which coincides with our solution to the primal program. In other words, there is no duality gap. If one had wanted to solve only the dual problem leading to $\lambda = -1$, one would check the primal feasibility of $(x(-1), y(-1)) = (4, 2)$. Once this is established, the next step is to check whether the two objective functions coincide, namely to make sure that no duality gap exists. Once this is established, we have solved also the primal problem.

Finally, note that Theorem 4.7 is applicable here. Specifically, for $\lambda^* = 1$,

$$L(x, y, -1) = -xy + (x - 3)^2 + y^2 - 5.$$

Its Hessian equals

$$\begin{pmatrix} 2 & -1 \\ -1 & 2 \end{pmatrix},$$

which is positive. Also, $(x^*, y^*) = (4, 2)$ is primal feasible, and it meets the FOCs.

Example 2 (Duality in linear programming). A standard linear program (LP) is defined as

$$\min_{x \in R^n} c^T x, \tag{4.12}$$

$$\text{s.t. } Ax = b,$$

$$x \geq \underline{0},$$

where the input is $c \in R^n$ (the price vector), $b \in R^m$ (the right-hand side), and $A \in R^{m \times n}$ (the constraint matrix). Linear programs are an important field in optimization, and Part III of this book is devoted to it.

The dual problem, in which only equality constraints are used when the dual objective function is defined, is

$$\phi(\lambda) = \min_{x \geq \underline{0}} \left\{ c^T x - \lambda^T (Ax - b) \right\},$$

where $\lambda \in R^m$. Clearly,

$$\phi(\lambda) = \min_{x \geq \underline{0}} \left\{ (c^T - \lambda^T A)x \right\} + \lambda^T b.$$

It is easy to see that if one (or more) of the entries in $c^T - \lambda^T A \in R^n$ are negative, we can make the corresponding entry in x as large as possible, making $\phi(\lambda)$ as negative as we wish. On the other hand, if all of the entries of $c^T - \lambda^T A \in R^n$ are non-negative, our optimal choice for x is $x = \underline{0}$. Thus,

$$\phi(\lambda) = \begin{cases} \lambda^T b, & c^T - \lambda^T A \geq \underline{0}, \\ -\infty, & \text{otherwise.} \end{cases}$$

Hence, the dual problem is

$$\max_{\lambda \in R^m} \phi(\lambda) = \max_{\lambda \in R^m} \{\lambda^T b \mid \text{s.t. } c^T - \lambda^T A \geq \underline{0}\}.$$

Note that if there is no $\lambda \in R^m$, which is dual feasible, namely there is no λ which satisfies $c^T - \lambda^T A \geq \underline{0}$, the above maximization returns $-\infty$.

To summarize, we get that the dual problem is the following linear program (LP):

$$\max_{\lambda \in R^m} b^T \lambda, \tag{4.13}$$

$$\text{s.t. } \lambda^T A \leq c^T.$$

This is the famous dual linear problem of the now called primal standard LP. In the final remark in Section 7.3.4 below, we show that no duality gap exists here. It is possible to see that, in this special case, the dual problem of the dual problem is the primal LP.[i] This analysis leads to the dual theorem of linear programming which says the following:

1. An LP has an optimal solution if and only its dual LP has an optimal solution. Moreover, in this case the optimal objective values coincide.
2. Any feasible solution of one of them yields a bound on any feasible objective function value of the other. In particular, if one of the two LPs (the primal or the dual) is unbounded, then the other is infeasible.
3. If one of the two LPs is infeasible, then the other is either infeasible or unbounded.

Example 3 (Duality in quadratic programming). Consider the following quadratic constrained optimization problem:

$$\min_{x \in R^n} \frac{1}{2} x^T Q x + c^T x,$$

$$\text{s.t } Ax = b,$$

for some symmetric positive matrix $Q \in R^{n \times n}$, some vector $c \in R^n$, some full-rank matrix $A \in R^{m \times n}$, and some vector $b \in R^m$. Assume that $m \leq n$. Note that the full-rank assumption implies that the problem is feasible. This, coupled with the fact that the relaxed unconstrained problem is bounded, implies that the problem we face is bounded too. As in

---

[i]To see this, one needs to convert the current dual LP into an equivalent standard LP and then take the dual of the latter. For details see, e.g., [15].

Section 2.1.2, we assume, without loss of generality, that $c = \underline{0}$. Moreover, due to another change of variables ($y = Q^{1/2}x$), we can, but do not, assume without loss of generality that $Q = I$, and hence this problem was already treated and solved as Example 1 in Section 3.4. In what follows, we solve it again using the duality theorem.

The dual function is

$$\phi(\lambda) = \min_{x \in R^n} \left\{ \frac{1}{2} x^T Q x - \lambda^T (Ax - b) \right\},$$

where $\lambda \in R^m$. Deriving the value of the dual function is an easy task as it is an unconstrained quadratic optimization function. In particular, from (2.9), it follows that the minimization defining the dual function is attained at

$$x^*(\lambda) = Q^{-1} A^T \lambda, \tag{4.14}$$

and hence, by minimal algebra, we get that

$$\phi(\lambda) = -\frac{1}{2} \lambda^T A Q^{-1} A^T \lambda + \lambda^T b.$$

Using (2.9) again and noticing that we are solving a maximization problem, namely minimizing $-\phi(\lambda)$, we get that

$$\lambda^* = \arg \max_{\lambda \in R^m} \phi(\lambda) = (A Q^{-1} A^T)^{-1} b,$$

and

$$\phi(\lambda^*) = \frac{1}{2} b^T (A Q^{-1} A^T)^{-1} b,$$

which is the value of the optimal objective function (of both the primal and dual problems). Finally, the optimal solution of the primal problem is, by (4.14),

$$x(\lambda^*) = Q^{-1} A^T \lambda^* = Q^{-1} A^T (A Q^{-1} A^T)^{-1} b.$$

It is easy to check that $x(\lambda^*)$ is primal feasible, namely $Ax(\lambda^*) = b$ and that the value of its objective function equals

$$\frac{1}{2} x^T (\lambda^*) Q x(\lambda^*) = \frac{1}{2} b^T (A Q^{-1} A^T)^{-1} b.$$

In other words, there is no duality gap.

4.6 Appendix: The Implicit Function Theorem

Consider all the points (x, y) on the unit circle. They obey the condition

$$x^2 + y^2 = 1.$$

Is y a function of x? The first answer that comes to mind is "no":

$$y = \pm\sqrt{1 - x^2}, \quad -1 \leq x \leq 1.$$

The above is not a function. Yet, if we take a point that meets the condition, say $(\frac{\sqrt{2}}{2}, \frac{\sqrt{2}}{2})$, and look for the continuous and differentiable function y of x that crosses this point along some (maybe small) interval around $x = \frac{\sqrt{2}}{2}$ (where this point is in the interior of the interval), then, the function is

$$y = \sqrt{1 - x^2}, \quad \frac{\sqrt{2}}{2} - \epsilon \leq x \leq \frac{\sqrt{2}}{2} + \epsilon$$

for some (not necessarily small) $\epsilon > 0$. Had the point been $(\frac{\sqrt{2}}{2}, -\frac{\sqrt{2}}{2})$, the function would then have been

$$y = -\sqrt{1 - x^2}, \quad \frac{\sqrt{2}}{2} - \epsilon \leq x \leq \frac{\sqrt{2}}{2} + \epsilon.$$

See Figure 4.3. However, if we take the point $(1, 0)$ (or $(-1, 0)$) that meets the condition $x^2 + y^2 = 1$, this task of finding the function becomes impossible. Indeed, decreasing $x = 1$ by a bit (the only feasible incremental change) leads to an ambiguity in the definition of y, making y not a single-value function of x, as required from a function.

What went wrong in $(1, 0)$ as a choice for the point around which a function can be defined? If we look at the derivatives of the above two functions $y = \pm\sqrt{1 - x^2}$, we observe that when $x \neq 1$, the derivatives are

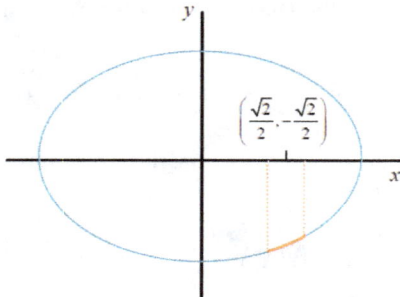

Figure 4.3. Unit circle.

well defined and finite; they are $-\frac{x}{\sqrt{1-x^2}}$ and $\frac{x}{\sqrt{1-x^2}}$. However, when $x = 1$, the corresponding derivatives do not exist.

The explicit function theorem says that y is an (implicit) differentiable (and hence continuous) function of x that is determined by the condition $x^2 + y^2 = 1$ as long as some feasible and *regular* point of it is given, and this is the case for some small enough neighborhood around the given point. The above definition of regularity is somewhat vague. The formal theorem, known as the *implicit function theorem,* is stated next.

Theorem

Theorem 4.8. *Let $g_i(x)$, $1 \leq i \leq m$, be m differentiable functions from R^n to R^1, with $m < n$. Consider a point $x^0 \in R^n$ with $g_i(x^0) = 0$, $1 \leq i \leq m$. Let $J \in R^{m \times m}$ be the Jacobian matrix such that its ith row contains the first m partial derivatives with respect to $x_1, \ldots, x_m$ of $g_i(x)$ at the point x^0, $1 \leq i \leq m$. If J is invertible, then there exists a neighborhood in R^{n-m} around $\hat{x}^0 = (x^0_{m+1}, \ldots, x^0_n)$ and m differentiable functions $\phi_i(\hat{x}) : R^{n-m} \to R$, $1 \leq i \leq m$, such that*

$$\phi_i(\hat{x}^0) = x^0_i, \quad 1 \leq i \leq m,$$

and

$$g_i(\phi_1(\hat{x}), \ldots, \phi_m(\hat{x}), \hat{x}_{m+1}, \ldots, \hat{x}_n) = 0, \quad 1 \leq i \leq m.$$

Exercise 4.6

Consider the following constrained minimization problem:

$$\min \frac{1}{2}||x||^2,$$

$$\text{s.t.} \quad Ax = b,$$

where $x \in R^n$, and $A \in R^{m \times n}$ is a full-rank matrix with $m < n$.

(a) Solve this problem while stating the KKT FOCs. In particular, find both the optimal solution and the optimal objective value. Are the FOCs sufficient? Why?
(b) Formulate the dual function as an unconstrained minimization problem.
(c) Solve the optimization problem mentioned above.
(d) Formulate the dual optimization problem.
(e) Solve the dual optimization problem.

(f) Compare the objective values of both the primal and dual problems. Do they coincide?

Exercise 4.7

Consider the following optimization problem:

$$\max_{x_1,x_2,x_3} x_1 x_2 + x_1 x_3 + x_2 x_3, \quad \text{s.t.} \quad x_1 + x_2 + x_3 = 3.$$

(a) Convert the problem of finding a stationary point into solving a system of equations.
(b) Replace the 3 on the right-hand side with b. For any b in the neighborhood of 3, derive $x_i^*(b), 1 \le i \le 3$, and the optimal value as a function of b. What is the derivative of this function when $b = 3$?

Exercise 4.8

(a) Solve

$$\min_x x^T Q x, \quad \text{s.t.} \quad Ax = b,$$

where Q is positive, and A is a full-rank matrix with more columns than rows.
(b) Solve item (a) for the special case where $Q = \begin{bmatrix} 2 & 1 \\ 1 & 2 \end{bmatrix}$, $A = (1, 2)$, and $b = 1$.
(c) Show that the solution to

$$\min_x \left(x^T Q x + s(Ax - b)^T (Ax - b) \right),$$

for any positive scalar s is:

$$x = \left(\frac{Q}{s} + A^T A \right)^{-1} A^T b.$$

(d) Solve the problem in item (c) while using the result stated item (b).
(e) Does the solution of (d) in the limit $s \to \infty$ coincide with that of (b)? Prove your answer and explain why.
(f) What is wrong with simply replacing $\frac{Q}{s}$ with zero when we take the limit in the answer of (c)?

Exercise 4.9

Consider the following optimization problem (it appears in [3, pp. 212–213]):

$$\min_{x_1, x_2, x_3} \quad 2x_1 + 3x_2 - x_3,$$

$$\text{s.t.} \quad x_1^2 + x_2^2 + x_3^2 = 1,$$

$$x_1^2 + 2x_2^2 + 2x_3^2 = 2.$$

(a) Write the KKT set of equations.
(b) Show that this set of equations does not have a solution.
(c) Deduce that the optimal solution is irregular, and hence in the optimal solution, the gradients of the constraints are proportional.
(d) Show that the previous item implies that the optimal solution satisfies $x_1^* = 0$ or $x_2^* = x_3^* = 0$.
(e) Rule out the second option in item (d) and conclude that $x_1^* = 0$.
(f) Solve the new problem stemming from the assumption that $x_1^* = 0$.
(g) Prove that the optimal point is $(0, -3/\sqrt{10}, 1/\sqrt{10})$.
(h) What is the worst point (i.e., where maximization is attained)?

Exercise 4.10

Consider the following (primal) optimization problem:

$$\min_{x_1, x_2} \{x_1^2 + x_2^2 + 2x_1\},$$

$$\text{s.t.} \quad x_1 + x_2 = 0.$$

(a) Prove that the objective function is convex, and the feasible set is convex.
(b) Solve this optimization problem by utilizing a single Lagrange multiplier.
(c) Show that the two sufficient regulation conditions stated in Section 4.2 hold here.
(d) State the dual function and the dual optimization problem.
(e) Solve the dual problem. In particular, show that the optimal dual solution coincides with the value of the Lagrange multiplier of the primal problem, and the optimal values of the primal and dual problems coincide.

Chapter 5

Optimization Under Equality and Inequality Constraints

5.1 Introduction

Let us take a bird's-eye view of the issue of the optimization under equality constraints. Clearly, each added constraint makes the problem more restrictive, and any given point needs to meet more conditions in order to be feasible. Yet, once a point is feasible, it has less competition for optimality. This is due to the fact that those previously feasible points that do not meet the new constraints are now not feasible and hence are eliminated from the set of feasible points. Note that when constraints are added, the gradient of the objective function needs to lie in a larger dimension linear subspace that contains the previous gradients: The gradient of the added constraint joins the gradients of the previous constraints to form a basis. Hence, it is more likely that the gradient of the point under consideration lies in this larger linear subspace.

Suppose now that an inequality constraint $h(x) \leq 0$ is added, and suppose that a point x obeys this and the previous constraints. Does this help us in terms of increasing the subspace in which $\nabla f(x)$ needs to lie at? In order to answer this question, we need to consider two exhaustive and mutually exclusive cases: $h(x) < 0$, when the added constraint is not binding (sometimes is referred to as an inactive constraint) at x, and the case where $h(x) = 0$, when the constraint is binding (or active) at x.

Let us deal with the first case. Given that the constraint is not binding, then changing x a bit in any direction does not lead to a violation of this constraint. Hence, the set of directions we need to consider in order to check for local optimality at x is not reduced under this additional constraint. Put differently, $\nabla h(x)$ cannot be added to the basis that spans the linear subspace in which $\nabla f(x)$ needs to belong to. It may look odd to mention this, but this is the equivalent of insisting on adding $\nabla h(x)$ to the basis, while forcing the coefficient of $\nabla h(x)$ in $\nabla f(x)$'s expression of a linear combination of said basis, to equal zero. Put differently, this is equivalent to imposing the Lagrange multiplier of $\nabla h(x)$ to equal zero. The question of looking for Lagrange multipliers is now back to the case where only the original equality constraints are considered.

Now consider the second case, namely where $h(x) = 0$. First, add this constraint to the previous set of constraint but now as an equality constraint. Note that if x is optimal for the case with the constraint $h(x) \leq 0$, then it is also optimal for the case where the constraint is $h(x) = 0$. In particular, for optimality, it should meet the FOCs for the case of only equality constraints. Finally, and this is the crucial part, consider the sign of the corresponding Lagrange multiplier, denoted by μ. Recall from Section 4.4 that μ is the derivative of the optimal objective function with respect to the corresponding entry on the right-hand side (currently at zero). Thus, if we decrease this entry by an infinitesimal $\epsilon > 0$, the optimal value will go down by $\mu\epsilon$. If $\mu > 0$, then from the optimality point of view, it would be worthwhile to make this decrease. Note that given the inequality constraint optimization we face, this decrease is possible. Hence, in order not to make it worthwhile, namely still having a chance that x is optimal, we need $\mu \leq 0$. Note that a possible way to combine the two cases is to require that $\mu \cdot h(x) = 0$ and $\mu \leq 0$ (on top of $h(x) \leq 0$).

Finally, note that in the case where the maximization of an objective function is considered, all of the above applies verbatim with one (crucial) exception: It must be the case that $\mu \geq 0$.

5.2 Optimization Under Equality and Inequality Constraints

The following theorem was just argued for and is the KKT conditions version of the constrained optimization studied in this section.

Theorem

Theorem 5.1. *Consider the following constrained optimization problem:*

$$\min_x f(x), \tag{5.1}$$

$$s.t. \quad g_i(x) = 0, \quad 1 \leq i \leq m,$$
$$h_j(x) \leq 0, \quad 1 \leq j \leq k.$$

Assume that x^* is a feasible solution. Also, assume that the gradients of the binding constraints are linearly independent at x^*.[a] Then, a necessary condition for x^* to be a local minimum is the existence of λ_i, $1 \leq i \leq m$, and $\mu_j \leq 0$, $1 \leq j \leq k$, such that

$$\nabla f(x^*) - \sum_{i=1}^{m} \lambda_i \nabla g_i(x^*) - \sum_{j=1}^{k} \mu_j \nabla h_j(x^*) = \underline{0} \in R^n$$

and

$$\sum_{j=1}^{k} \mu_j h_j(x^*) = 0.$$

---

[a]The equality constraints are binding by definition.

Note that the condition $\mu_j h_j(x^*) = 0$, $1 \leq j \leq k$, follows logically from the conditions stated in the above theorem. These are known as the *complementary slackness conditions*. In particular, the Lagrange multipliers associated with the non-binding constraints are zero. On the other end and although unlikely, note that zero is a possible value for the Lagrange multiplier for a binding constraint.

Note that the generalization of Theorem 4.5 holds verbatim here. In particular, as expected, the derivative of the objective function with respect to the the right-hand side entry of a non-binding constraint equals zero, which is the value of the corresponding Lagrange multiplier.

Example. Consider the following convex optimization problem:

$$\min_{x_1, x_2} \quad 4x_1^2 + x_2^2 - x_1 - 2x_2,$$
$$s.t. \quad 2x_1 + x_2 \leq 1,$$
$$x_1^2 \leq 1.$$

Solution. Introducing the set of Lagrange multipliers, we get that the Lagrangian function is

$$L(x_1, x_2, \mu_1, \mu_2) = 4x_1^2 + x_2^2 - x_1 - 2x_2 - \mu_1(2x_1 + x_2 - 1) - \mu_2(x_1^2 - 1).$$

The KKT conditions are:

(i) $8x_1 - 1 + 2\mu_1 - 2\mu_2 x_1 = 0,$
(ii) $2x_2 - 2 - \mu_1 = 0,$
(iii) $\mu_1(2x_1 + x_2 - 1) = 0,$
(iv) $\mu_2(x_1^2 - 1) = 0,$
(v) $2x_1 + x_2 \leq 1,$
(vi) $x_1^2 \leq 1,$
(vii) $\mu_1, \mu_2 \leq 0.$

We first start with the guess $\mu_1 = 0$. Then (ii) implies that $x_2 = 1$, while (i) yields $x_1 = \frac{1}{8 - 2\mu_2}$. Since $\mu_2 \leq 0$, then, of course, $x_1 < 1$, and therefore, from (iv), we conclude that $\mu_2 = 0$, which leads to $x_1 = \frac{1}{8}$. This result, coupled with $x_2 = 1$, contradicts (v), thereby overriding the assumption that $\mu_1 = 0$. Since the only option left is that $\mu_1 < 0$, it follows from (iii), that $2x_1 + x_2 - 1 = 0$, namely, $x_2 = 1 - 2x_1$. We can then transform our original problem to the following single-variable problem:

$$\min_{x_1} \quad 4x_1^2 + (1 - 2x_1)^2 - x_1 - 2(1 + 2x_1),$$
$$\text{s.t.} \quad x_1^2 \leq 1.$$

The objective function is quadratic with a global minimum at $x_1 = \frac{1}{16}$. As it turns out, this solution satisfies the constraint $x_1 \leq 1$ (as a strict inequality), and hence it is the optimal solution to the constrained optimization problem. A simple calculation leads to the solution $(x_1, x_2) = (\frac{1}{16}, \frac{7}{8})$. For completeness, note that by (ii), $\mu_1 = -1/4$, and by (iv) (since $x_1 < 1$), $\mu_2 = 0$.

Exercise 5.1

Consider the following optimization problem:

$$\min_{x_1, x_2} x_1^2 + 2x_2^2 + 4x_1 x_2, \quad \text{s.t.} \quad x_1 + x_2 = 1, \quad x_1, x_2 \geq 0.$$

Convert the problem of finding a stationary point into one of solving a system of equations.

Exercise 5.2

Consider the following optimization problem:

$$\min_{x \in R^n} F(x),$$

where

$$F(x) = \min_{i=1}^{m} f_i(x),$$

for some m functions $f_i(x) : R^n \to R$, $1 \leq i \leq m$.

1. Formulate this problem as a constrained optimization problem.
2. What are the FOCs which are met by an optimal solution?
3. Why are no regularity conditions needed to be assumed in the case where two of the m functions never agree for the same x?

5.2.1 *Second-order conditions (SOCs)*

The following theorem states the second-order conditions (SOCs) for optimality.

Theorem

Theorem 5.2. *Let x^*, λ, and μ be a triple that meets all the conditions of Theorem 5.1. For the feasible solution x^*, let A be the (possibly empty) set of binding constraints out of the set of inequality constraints. Then, a necessary condition for local minimum at x^* is*

$$d^T \left(\nabla^2 f(x^*) - \sum_{i=1}^{m} \lambda_i \nabla^2 g_i(x^*) - \sum_{j \in A} \mu_j \nabla^2 h_j(x^*) \right) d \geq 0, \quad (5.2)$$

for all directions d that obey $d^T \nabla g_i(x^*) = 0$, $1 \leq i \leq m$, and $d^T \nabla h_j(x^*) = 0$, $j \in A$. Replacing the "$\geq$" sign above with "$>$" (but now $d \neq 0$) makes this condition sufficient.

5.2.2 *Convex programming*

Consider the following constrained optimization problem:

$$\min_x f(x),$$

$$\text{s.t.} \quad g_i(x) = 0, \quad 1 \leq i \leq m,$$

$$h_j(x) \leq 0, \quad 1 \leq j \leq k.$$

In order to construct a convex feasible set, two sufficient assumptions are needed. The first is that the functions $g_i(x)$, $1 \leq i \leq m$, are affine. Indeed, only in affine functions does $g_i(x^1) = g_i(x^2) = 0$ imply that $g_i(\alpha x^1 + (1 - \alpha)x^2) = 0$ for any α, and in particular for any α with $0 \leq \alpha \leq 1$. Note that for such functions, $\nabla^2 g_i(x) = \underline{0}$. The second assumption needed is that the functions $h_j(x)$, $1 \leq j \leq k$, are convex. In particular, $\nabla^2 h_j(x)$ is non-negative for any x, $1 \leq j \leq k$. Indeed, for such functions, if $h_j(x^1) \leq 0$ and $h_j(x^2) \leq 0$, then for any α with $0 \leq \alpha \leq 1$, $h_j(\alpha x^1 + (1 - \alpha)x^2) \leq \alpha h_j(x^1) + (1 - \alpha)h_j(x^2) \leq \alpha 0 + (1 - \alpha)0 = 0$.

Consider now Theorem 5.2. It is straightforward that if $f(x)$ is convex, then once x^*, coupled with the appropriate λ and μ, meets the FOCs for a minimum point, and the above posed two convexity conditions are met, then the SOCs are met as well. Thus, the existence of these two convexity conditions results in a much simple analysis: Only the FOCs need to be considered when optimization is concerned. More importantly, the counterpart of Theorem 4.4, which is stated next, holds too. As the proof is analogous to that of Theorem 4.4, it is omitted.

Theorem

Theorem 5.3. *If $f(x)$ is convex, $g_i(x)$ is affine, $1 \leq i \leq m$, and $h_i(x)$ is convex, $1 \leq i \leq k$, then, if x^* meets the FOCs for (5.1), x^* is a globally optimal solution.*

The branch of optimization theory applied to such problems is called *convex programming*. The terminology used in this section is particulary revealing, since the intersection of convex sets is itself a convex set, making the feasible set of a convex problem a convex set. The dual function is now clearly defined by $\phi(\lambda, \mu)$ as before, but with an additional constraint that needs to be put: $\mu \leq \underline{0}$. Moreover, in the case of convex programming, meeting the conditions specified in Theorem 5.2 is sufficient for global minimization (and not only local minimization).

Note that in the case of a maximization problem, the assumption of convex programming still requires $h_j(x)$, $1 \leq j \leq k$, to be convex (not concave, of course). Yet, $f(x)$ is assumed to be concave. Recall that now $\mu_j \geq 0$, $1 \leq j \leq k$, and since $\nabla^2 f(x^*)$ is non-positive, inequality (5.2) needs to be reversed.

Finally, we would like to claim that Example 1 through Example 7 in Section 4.2.1 meet the conditions for convex programming. For some of these examples, this claim was already proved above. The rest of the remaining examples are left as exercises for the reader. Finally, the same condition holds for the least squares line problem presented in Section 2.3.2. Specifically, for A as defined in that section, $\nabla^2 f(x^*) = A^T A$, a matrix that was shown to be invertible when A is a full-rank matrix. As such, the matrix is also positive.[a]

Remark on duality. The duality analysis done in the previous chapter applies verbatim to the problem dealt with here, where the set of Lagrange multipliers associated with the inequality constraints is just added to the set of Lagrange multipliers used there. The only technical requirement is the existence of a feasible point (and hence points) such that all inequality constraints are not nonbinding at this point. This is known as Slater's condition.

5.2.3 Example: MLEs for two independent categorical random variables

Suppose there are two categorical random variables. The first one is with k possibilities, while the second with h. Each individual in the population possesses one of the features from each of the two categories. Denote by α_i the probability to belong to option i at the first variable, $1 \leq i \leq k$, $\alpha_i \geq 0$, with $\Sigma_{i=1}^k \alpha_i = 1$. Define β_j, $1 \leq j \leq h$, in the same way for the second category. Finally, assume the two categories are independent.

In order to estimate the above defined probabilities, a sample of size N is taken. Denote by N_{ij} the random number of those that belong to the so-called (i, j)-cell, $1 \leq i \leq k$, $1 \leq j \leq h$. Assume that none of the values $\Sigma_{j=1}^h N_{ij}$, $1 \leq i \leq k$ and $\Sigma_{i=1}^k N_{ij}$, $1 \leq j \leq k$, equal zero. Up to a constant, which is not a function of the probability, the likelihood function is proportional to

$$\prod_{1 \leq i \leq k, 1 \leq j \leq h} (\alpha_i \beta_j)^{N_{ij}}.$$

The MLE maximizes this value. Alternatively, one can maximize the log of this function. Thus, one faces the following concave constraint

---

[a]**Proof.** $x^T A^T A x = \|Ax\| \geq 0$. Moreover, in the full-rank case, this inequality stands as equality if and only if $x = \underline{x}$.

maximization problem:

$$\max_{p_{ij}, 1 \leq i \leq k, 1 \leq j \leq h} \sum_{1 \leq i \leq k, 1 \leq j \leq h} N_{ij}(\log \alpha_i + \log \beta_j),$$

$$\text{s.t.} \quad \sum_{i=1}^{k} \alpha_i = 1,$$

$$\sum_{j=1}^{h} \beta_j = 1,$$

$$\alpha_i \geq 0, \quad 1 \leq i \leq k,$$

$$\beta_j \geq 0, \quad 1 \leq j \leq h.$$

This is a convex program which we solve next.

Since a zero value for any of the decision variables leads to minus infinity as the objective function value, the corresponding Lagrange values are zero. Thus, the Lagrangian function equals, in fact,

$$\sum_{1 \leq i \leq k, 1 \leq j \leq h} N_{ij}(\log \alpha_i + \log \beta_j) - \lambda_1 \left(\sum_{i=1}^{k} \alpha_i - 1 \right) - \lambda_2 \left(\sum_{j=1}^{h} \beta_j - 1 \right).$$

Taking the derivative with respect to α_i, we get

$$\frac{\sum_{j=1}^{h} N_{ij}}{\alpha_i} - \lambda_1, \quad 1 \leq i \leq k.$$

Setting it to zero, we get that

$$\alpha_i \propto \sum_{j=1}^{h} N_{ij}, \quad 1 \leq i \leq k.$$

The fact that $\Sigma_{i=1}^{k} \alpha_i = 1$, implies that the MLE for α_i equals

$$\frac{\sum_{j=1}^{h} N_{ij}}{N}, \quad 1 \leq i \leq k.$$

A similar expression exists for the MLE for β_j, $1 \leq j \leq h$.

Finally, inspecting again the above problem, we can see that we have a separable problem in the sense that the α and the β variables do not interact. Hence, they can be solved as two separate problems. Moreover, as it turns out, each one of them is in fact the same as the problem stated and solved in Section 3.2.4. In particular, the fact that the above result extends to the case where one of the values $\Sigma_{j=1}^{h} N_{ij}$, $1 \leq i \leq k$ and $\Sigma_{i=1}^{k} N_{ij}$, $1 \leq j \leq k$, equals zero is argued for there.

5.2.4 *Example: support vector machine (SVM)*

Consider the hyperplane that consists of all $x \in R^n$ that obey $a^T x - b = 0$. Let $x_i \in R^n$, $1 \le i \le m$, be a set of m points. The point i is said to be above (respectively, below) the hyperplane if $a^T x_i - b > 0$ (respectively, $a^T x_i - b < 0$), $1 \le i \le m$. In the former case, let $y_i = 1$, and in the latter, let $y_i = -1$, $1 \le i \le m$. Reversing the search, suppose that the values of y_i, $1 \le i \le m$, are given, and one then looks for such a hyperplane. In most cases, namely when the data is random, such a hyperplane does not exist. However, it does exist when the m points can be decomposed into two classes. In such cases, the hyperplane separates one class of points from the other. Admittedly, this is a tautology. See Figure 5.1 for the case where $n = 2$.

Once such a hyperplane exists, it is normal that many, in fact infinitely many, such hyperplanes exist. The focus is then on the one that separates them the most in a sense that will be made precise shortly. Of course, this concept calls for a definition, but first, note that $a^T x_i - b \ge C_1$ for all those points in the first class, for some $C_1 > 0$. Likewise, $a^T x_i - b \le -C_2$ for some $C_2 > 0$ for all those points in the second class. This can be stated as $y_i(a^T x_i - b) \ge \min\{C_1, C_2\}$, $1 \le i \le m$. Without loss of generality, we can replace the right-hand side by 1 (otherwise, we divide all entries in a and b by the right-hand side value without changing the hyperplane). Minding this constant, observe that in fact we get two parallel hyperplanes. One is $a^T x - b - 1 = 0$, where all points from the first class are on or above it. The other hyperplane is $a^T x - b + 1 = 0$, where all points from the second class are on or below it.

Our objective is to find the pair of $a \in R^n$ and $b \in R^1$, namely the two hyperplanes, $a^T x + b - 1 = 0$ and $a^T x + b + 1$, separated by the greatest distance. From (3.23), we know that this distance equals $2/\|a\|$. Thus, we wish to minimize $\|a\|$. In summary, our objective is

$$\min_{a \in R^n, b \in R} \|a\|^2,$$

$$\text{s.t. } y_i(a^T x_i - b) \ge 1, \quad 1 \le i \le m.$$

Of course, each class of points has its closest point to the optimal hyperplane, and to some extent, these two points characterize the sets. All that is required from the other points is just to be on the side of their class. These two points are referred to as the *support vectors*. It is possible to argue that these two points obey the required inequality constraint as an equality. Indeed, as Figure 5.1 shows, were this not the case, it would be

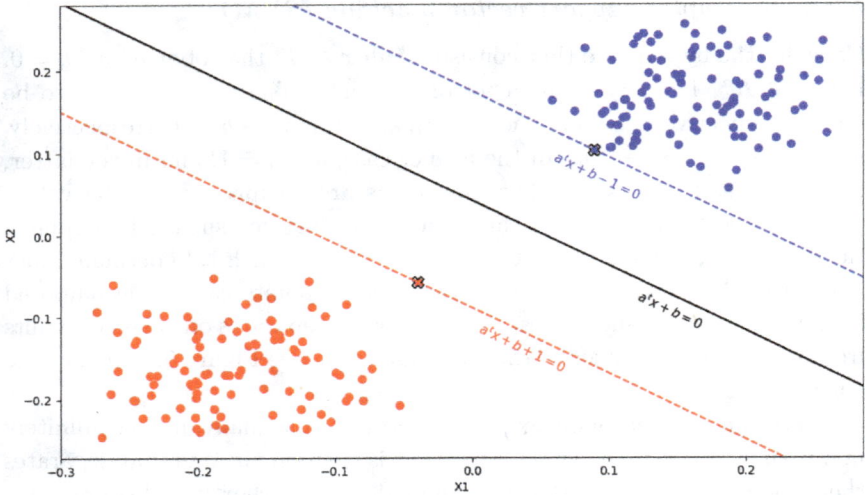

Figure 5.1. Two-dimensional SVM.

possible to increase the distance between the two hyperplanes $a^T x + b = 1$ and $a^T x + b = -1$. Finally, in the case where such a separating hyperplane does not exist for the given decomposition of the data into two classes, the above program will be infeasible. There exists a vast literature on how to model such data, but that would take us beyond the scope of this text.

Exercise 5.3

(a) Show that the best way to average a number of uncorrelated unbiased estimators is to give each one of them a weight that is proportional to the inverse of its variance.

(b) Show that the corresponding optimal variance is the harmonic mean of the variances of the individual estimators, divided by the number of estimators.

(c) Conclude that the harmonic mean is less than or equal to the arithmetic mean.

(d) Assume that the estimators are uncorrelated. That is, assume that there exists some variance-covariance matrix Σ that is not necessarily diagonal. What is the optimal estimator? What is the corresponding variance? Hint: The objective of this new mathematical program is to minimize a sum of squares (and to change the equality constraint accordingly).

Exercise 5.4

Consider the following program:

$$\min_{a,b} \int_{x=0}^{1} (x^2 - ax - b)^2 dx, \quad \text{s.t.} \quad a + b = 1.$$

(a) What are the first- and the second-order conditions for the local minimum point?
(b) Is this a convex program?
(c) What is the value of the optimal Lagrangian function and the value of the Lagrange multiplier?
(d) Suppose that the constraint $a + b = 1$ is replaced by one of the following constraints:

- $a + b = 1.01$,
- $a + b \leq 1$,
- $a + b \geq 1$.

Say in your own words and without doing any further analysis, what one can infer on these three cases from the original version of the program.

5.3 Algorithms for Optimization Under Constraints

5.3.1 *The first-order Lagrangian method*

Admittedly, the method presented here is an example of a penalty function approach that is the topic of the next section, but nevertheless, due to its importance, it is dealt with separately here. Consider the following problem:

$$\min_{x \in R^n} f(x),$$

$$\text{s.t.} \quad g_i(x) = 0, \quad 1 \leq i \leq m.$$

Recall the Lagrangian function $L(x, \lambda) : R^{n+m} \to R$ where $L(x, \lambda) = f(x) - \lambda^T g(x) = f(x) - \sum_{i=1}^{m} \lambda_i g_i(x)$. Next, define the (penalty) function

$$P(x, \lambda) = \frac{1}{2} \|\nabla_x L(x, \lambda)\|^2 + \frac{1}{2} \|g(x)\|^2.$$

It is easy to see that $P(x, \lambda) \geq 0$. Moreover, the equality holds if and only if x is feasible, and coupled with λ, it satisfies the FOCs for optimality. Thus, the equality-constrained optimization problem is reduced to an unconstrained optimization problem, albeit with $n + m$ variables. Moreover, if the optimal objective function value is zero, we then have in hand a global optimal solution.

Our goal is to solve $\min_{x \in R^n, \lambda \in R^m} P(x, \lambda)$. A straightforward approach is to apply one of the techniques introduced in Chapter 2 that uses iterative procedures for solving unconstrained optimization problems, such as the steepest descent and Newton methods. It is possible to see that

$$\nabla P(x, \lambda) = \begin{pmatrix} \nabla^2_{xx} L(x, \lambda) \nabla_x L(x, \lambda) + \nabla \underline{g}(x)^T g(x) \\ -\nabla \underline{g}(x) \nabla_x L(x, \lambda) \end{pmatrix}.$$

Thus, if (x_k, λ_k) is some current approximation for (x^*, λ^*), then if one applies the steepest descent method, the next approximation can be $(x_k, \lambda_k)^T - \alpha_k \nabla P(x_k, \lambda_k)^T$, where the scalar $\alpha_k > 0$ is found via some line search.

An alternative to the descent direction of $-\nabla P(x, \lambda)$ just used is the direction $-(\nabla_x L(x, \lambda), g(x))$. The reason behind this choice is that it is descent direction too. Indeed, it is possible to show that

$$-(\nabla_x L(x, \lambda), \underline{g}(x))^T \nabla P(x, \lambda) = -\nabla_x L(x, \lambda)^T \nabla^2_{xx} L(x, \lambda) \nabla_x L(x, \lambda).$$

Thus, if this expression is negative, $-(\nabla_x L(x, \lambda), g(x))$ is a descent direction at (x, λ). Clearly, a sufficient condition for this to be the case is that $\nabla^2_{xx} L(x, \lambda)$ is positive. Of course, this alternative direction is simpler to compute than $\nabla P(x, \lambda)$ is, as it does not involve second-order derivatives.

5.3.2 *Penalty functions*[b]

Consider the following constrained optimization problem:

$$\min_x f(x), \quad \text{s.t.} \quad x \in S. \tag{5.3}$$

Define some function $P(x) : S \rightarrow R$. In the present context, we call such a function a *penalty function* as it obeys the following requirements. Assume

---

[b]This section is based on Chapter 13 of [18].

that $P(x) = 0$ for $x \in S$ and $P(x) > 0$ for $x \notin S$. This function can be understood as a price, or penalty, that one pays due to violating the constraint. For example, suppose that:

$$\min_{x \in R^n} \quad f(x),$$

$$\text{s.t.} \quad h_i(x) \leq 0, \quad 1 \leq i \leq m.$$

Then, a possible penalty function can be

$$P(x) = \frac{1}{2} \sum_{i=1}^{m} (\max\{0, h_i(x)\})^2.$$

In words, the greater the violation of the constraint, the greater the penalty.

Let $C_k > 0$ be some number and consider the following unconstrained optimization problem:

$$\min_{x \in R^n} \{f(x) + C_k P(x)\}.$$

This of course is not the problem we wish to solve, but it is an unconstrained problem for which we have a few algorithms in our arsenal, such as the steepest-descent algorithm or Newton's method including all its variations. Without being too scientific, we can say that the greater the value of C_k is, the better, in the sense that one is the less likely to violate the constraints. This hypothesis is made formal and proved for correctness in the following theorem. Yet, when C_k is truly large, it may have numerical issues which are beyond the scope of this text.

Theorem

Theorem 5.4. *Consider the constrained optimization problem* (5.3) *and denote its optimal solution by* x^*. *Also, let* C_k, $1 \leq k < \infty$, *be a monotone increasing series of real numbers with* $\lim_{k \to \infty} C_k = \infty$. *Define the functions* $q(x, C_k) = f(x) + C_k P(x)$ *and the values* $x_k = \arg\min_x q(x, C_k)$, $k \geq 0$. *Then,*

1. $q(x_k, C_k) \leq q(x_{k+1}, C_{k+1})$, $k \geq 1$.
2. $P(x_k) \geq P(x_{k+1})$, $k \geq 1$.
3. $f(x_k) \leq f(x_{k+1})$, $k \geq 1$.
4. $f(x_k) \leq f(x^*)$, $k \geq 1$.
5. $\lim_{k \to \infty} x_k = x^*$.

Proof.

(1) By definition,

$$q(x_{k+1}, C_{k+1}) = f(x_{k+1}) + C_{k+1}P(x_{k+1}) \geq f(x_{k+1}) + C_k P(x_{k+1})$$
$$\geq f(x_k) + C_k P(x_k) = q(x_k, C_k), \quad k \geq 1,$$

where the second inequality follows from the optimality of x_k for $f(x) + C_k P(x)$.

(2) Note that

$$f(x_k) + C_k P(x_k) \leq f(x_{k+1}) + C_k P(x_{k+1}),$$

and

$$f(x_{k+1}) + C_{k+1}P(x_{k+1}) \leq f(x_k) + C_{k+1}P(x_k).$$

Summing up these two inequalities and performing some minimal algebra, we get that

$$(C_{k+1} - C_k)P(x_{k+1}) \leq (C_{k+1} - C_k)P(x_k).$$

The fact that $C_{k+1} - C_k > 0$ concludes the proof.

(3) Since $f(x_k) = q(x_k, C_k) - P(x_k)$ and since $f(x_{k+1}) = q(x_{k+1}, C_{k+1}) - P(x_{k+1})$, it follows from the first two items above that $f(x_k) \leq f(x_{k+1})$.

(4) Follows from Item 3 by the continuity of $f(x)$.

(5) See [18, p. 412]. $\square$

Admittedly, selecting a good penalty function may sometimes seem more as a form of an art than science. Next, we show a penalty function, which turns out to be ideal.

Example. Consider the following problem:

$$\min_{x,y} \ 2x^2 + 2xy + y^2 - 2y,$$

$$\text{s.t.} \quad x = 0.$$

First, note that the solution is trivially found to be $x = 0$ and $y = 1$. Second, it is easy to derive that $\lambda = 2$ is the value of the Lagrange multiplier of the constraint $x = 0$. Third, using the penalty function $P(x) = |x|$, we get that

$$q(x, y, C) = 2x^2 + 2xy + y^2 - 2y + C|x|.$$

Next, we argue that for $C > 2$, the minimization here is obtained in $(x, y) = (0, 1)$, indicating that the limit results stated in Theorem 5.4 can be reached after a finite number of steps, specifically when C is large enough. Indeed,

$$2x^2 + 2xy + y^2 - 2y + C|x| = x^2 + (2x + C|x|) + (y - 1 + x)^2 - 1.$$

Hence, no matter what the value for x is, $y = 1 - x$ is required for minimization. The minimization for x^2 is attained at $x = 0$ as is $2x + C|x|$ when $C > 2$.

Remark. In the above example, C was taken to be larger than $|\lambda| = 2$. This is not a coincidence: It is known that if the penalty function for the problem posed in (5.1) is

$$P(x) = \sum_{i=1}^{m} |g_i(x)| + \sum_{j=1}^{p} \max\{0, h_i(x)\},$$

then for any C with $C \geq \max\{|\lambda_i|, 1 \leq i \leq m, |\mu_j|, 1 \leq j \leq p\}$, it leads to the optimal solution. See [5, p. 506]. Thus, if somehow one can guess an upper bound on the Lagrange multipliers, a constrained optimization can be reduced to an unconstrained one. Finally, note that the unconstrained optimization problem solved here is not differential, which makes the use of standard techniques, for example, the steepest descent and Newton's, a bit more involved.

5.3.3   *Barrier functions*

Suppose that we wish to solve the following problem:

$$\min_{x} f(x),$$

$$\text{s.t.} \quad g_i(x) \leq 0, \quad 1 \leq i \leq m.$$

Due to reasons that will be apparent soon, we need to assume that there exists an interior point x in the sense that $g_i(x) < 0$, $1 \leq i \leq m$. This is, once again, Slater's condition.

Let $B(x) : R^n \to R$, be a function with the property that $B(x) = \infty$ if $g_i(x) = 0$ for at least one i, $1 \leq i \leq m$, while $B(x) < \infty$ if $g_i(x) < 0$ for all $1 \leq i \leq m$. A possible example is

$$B(x) = -\sum_{i=1}^{m} \ln(-g_i(x)).$$

The let $\epsilon_k > 0$, $k \geq 1$, with the property that $\lim_{k \to \infty} \epsilon_k = 0$. Then, define

$$x_k = \arg\min_x\{f(x) + \epsilon_k B(x)\},$$

where $\log(0)$ is defined as infinity. It is claimed that $\lim_{k \to \infty} x_k = x^*$. Indeed, if x^* is an interior point, then the barrier part $\epsilon_k B(x)$ is practically ignored, as its limiting contribution to the objective vanishes. On the other hand, if x^* lies in the the boundary of the feasible set, namely if x^* satisfies $g_i(x^*) = 0$ for some i, $1 \leq i \leq m$, then the fact that $\lim_{k \to \infty} \epsilon_k = 0$ mitigates the effect that $B(x_k)$ goes to infinity when k goes to infinity.

Example. Consider the following problem[c]:

$$\min_{x,y} \frac{1}{2}(x^2 + y^2),$$

$$\text{s.t.} \quad -x + 2 \leq 0.$$

Clearly, the optimal solution is $(x, y) = (2, 0)$. Note that the constraint is binding at the optimal point. If we solve this problem based on the barrier function suggested above, we get that

$$(x_k, y_k) = \arg\min_{x,y}\left\{\frac{1}{2}(x^2 + y^2) - \epsilon_k \ln(x - 2)\right\}, \quad k \geq 1.$$

Differentiating the above with respect to x and setting the derivative to zero, we get that

$$x_k - \epsilon_k \frac{1}{x - 2} = 0, \quad k \geq 1.$$

Doing the same with respect to y, we get $y = 0$. Solving this set of two equations, we get that $x_k = 1 + \sqrt{1 + \epsilon_k}$ and $y_k = 0$, $k \geq 1$. Taking limits, we see that $\lim_{k \to \infty}(x_k, y_k) = (2, 0)$, which coincides with the optimal point. Note that if, additionally, $\epsilon_k > 0$, $k \geq 1$, we will be "lucky" in the sense that the sequence of points generated in this way is feasible, namely $-x_k + 2 \leq 0$, $k \geq 1$. This allows us to abort the algorithm after a finite number of iterations, with a feasible point that well approximates the optimal solution.

---

[c]This example appears in [5, p. 449].

Remark. As in the case of the dual problem, one can define the penalty function only with respect to part of the constraints, while the optimization that is done iteratively will be a constrained one, minding the set of all of the other constraints. For this matter, some or all of these might be equality constraints.

Part III
Linear Programming

Chapter 6

Introduction and Examples

6.1 Linear Programming Formulation

It was already noted in Section 3.4.4 that equality-constrained optimization is rather trivial when both the objective function and the constraints are linear. This implies that a model with some depth, where all functions involved are linear, calls for some inequality constraints. We indeed impose the constraints that all decision variables are required to be non-negative. Thus, a standard linear program (LP), as was introduced in Example 2 in Section 4.5.2, is as follows:

$$\min_{x \in R^n} c^T x, \tag{6.1}$$

$$\text{s.t.} \quad Ax = b,$$

$$x \geq \underline{0},$$

where the input is $c \in R^n$ (the price vector), $b \in R^m$ (the right-hand side), and $A \in R^{m \times n}$ (the constraint matrix). Note that the choice of $\underline{0}$ is both without loss of generality and practical, due to many operations research models where this non-negativity constraint is a common feature. For example, see Chapter 3 in [23] for many examples of real-life optimization problems formulated as LPs.

Linear programming is an area of optimization in which one needs to optimize a linear function over a feasible set that is defined as a set of vectors that satisfy a finite number of linear constraints. The linear constraints are defined as the requirement that a linear function of the decision variables will be less than or equal to some given number. Note that strict inequalities are not allowed. By (6.1), we conclude that the model of standard LP is more restrictive in two dimensions:

1. Only equality-constrained optimization is allowed.
2. The decision variables need to be non-negative. Yet, we will argue below that these constraints are without loss of generality; in other words, any (non-standard) LP can be converted into a standard one.

Remark. Note that Ax is a linear combination of the columns of A: x_i is the coefficient of the ith column of A, $1 \leq i \leq n$. Hence, each of the n columns of A is associated with an entry in x and vice versa.

Definition

Definition 6.1. If the objective min above is replaced by max, and the equality sign $=$ above is replaced by $\leq$, then the LP is called a *canonical* LP.

Theorem

Theorem 6.1. *By adding variables and/or adding constraints, a standard LP can be formulated as a canonical one, and a canonical LP can be formulated as a standard one.*

An immediate conclusion from the above theorem is that an algorithm that is designed to solve one form of an LP is able to solve the other form.

Exercise 6.1

Prove Theorem 6.1.

Remark. The requirement of a standard LP is that, for a given decision variable, say x_i, $x_i \geq 0$. Yet, sometimes the postulated problem does not require this constraint, namely that x_i is not restricted in sign. Nevertheless, in order to apply an algorithm for solving a standard LP, it is required to transform a non-standard LP into a standard one. In the case dealt with here, this is done by introducing two decision variables, say x_i' and x_i'', such that x_i is replaced with $x_i' - x_i''$. Now, it is possible to see that one can add the constraints $x_i' \geq 0$ and $x_i'' \geq 0$. Any feasible solution to the new

LP has a corresponding solution in the original LP, where their objective function values coincide. On the other hand, any feasible solution to the original LP has many corresponding solutions in the new LP, sharing with it the same objective value. Moreover, it is possible to see that one, and only one, of these many corresponding solutions has at least one variable, x_i' or x_i'', with a zero value. Finally, both of its variables can equal zero if and only if $x_i = 0$.

As it turns out, such problems — linear programs — are relatively easily solved, and hence it is tempting to model real-life problems as LPs. Of course, the quality of this modeling varies with the application. This part of the book examines some of the algebraic and geometric properties of the feasible set of solutions to a linear program, states the simplex method, and proves its finite convergence to the optimal solution (when it exists).

6.2 Examples: Not Least Squares Regression

The case of least squares regression was already studied in Section 2.3.3. The goal there was to find for $A \in R^{n \times m}$ and $b \in R^n$, where $n > m$, the solution $x^* \in R^m$ where

$$x^* = \arg \min_{x \in R^m} \sum_{i=1}^n \left(b_i - \sum_{j=1}^m A_{ij} x_j \right)^2 .$$

This is also known as the ℓ_2-norm regression, where the goal is to minimize the sum of squares of the residuals. The most commonly studied case is where $m = 2$ and $A_{i1} = 1$, $1 \le i \le n$. Here, one looks for the so-called simple linear regression line which was dealt with at length in Section 2.3.3, where a line is fitted to present points, (A_{i2}, b_i), $1 \le i \le n$. This set of points is usually referred to as (x_i, y_i), $1 \le i \le n$. The solution is given in (2.17) and (2.18) above. Finally, in the trivial case where $m = 1$ and $A_{i1} = 1$, $1 \le i \le n$, the optimal x^* is the arithmetic mean value across the entries b_i, $1 \le i \le n$. See Example 2 in Section 1.2.

We now turn our attention to two different objective functions.

6.2.1 *Minimizing the sum of the absolute residuals*

The first out of the two objective functions we consider is

$$\min_{x \in R^m} \sum_{i=1}^n \left| b_i - \sum_{j=1}^m A_{ij} x_j \right| , \tag{6.2}$$

which is known as the ℓ_1-norm regression, where the goal is to minimize the sum of the absolute values of the residuals. In the case where $m = 1$ and $A_{i1} = 1, 1 \le i \le n$,

$$x^* = \arg\min_{x \in R} \sum_{i=1}^{n} |b_i - x| \qquad (6.3)$$

is the median of the entries b_i, $1 \le i \le n$. This point is exemplified next.

Example. Suppose that the following observations are given by $b = \{-1, 0, 2, 4, 7\}$. The median of this five-entry series, (b), is of course 2. We will show that the median minimizes the loss function defined in (6.3). For any real number x, we get that the function which sums up the absolutes residuals of the given five-entry series is

$$h(x) = |-1 - x| + |0 - x| + |2 - x| + |4 - x| + |7 - x|$$

$$= \begin{cases} -5x + 12, & \text{for } x \le -1 \\ -3x + 14, & \text{for } -1 \le x \le 0 \\ -x + 14, & \text{for } 0 \le x \le 2 \\ x + 10, & \text{for } 2 \le x \le 4 \\ 3x + 2, & \text{for } 4 \le x \le 7 \\ 5x - 12, & \text{for } 7 \le x. \end{cases}$$

Let us explain: For $x \le -1$, all the expressions within the absolute value are positive, so in the function, these absolute value signs can be disregarded, and then the function is defined as $h(x) = -5x + 12$. When $-1 \le x \le 0$, this fact is true only for the last four expressions. For the first expression, the absolute value becomes the opposite sign of the expression inside it, and hence, the first expression equals $x + 1$. So, for this domain where $-1 \le x \le 0$, the function is defined as $h(x) = -3x + 14$. In the range $0 \le x \le 2$, the first two expressions change their sign, and the last three do not, so for this domain, the function is set to $h(x) = -x + 10$, and so on for other ranges of x. The graph for this function is given in Figure 6.1.

You can see that for $x < 2$, the function is monotone decreasing, and for $2 \le x$, it is monotone increasing. In particular, the point $x = 2$ is a minimum point where the function changes the direction of monotonicity. Recall that $med(b) = 2$. In other words, the loss function that we defined is minimized at the median point. In the case where there is an even number

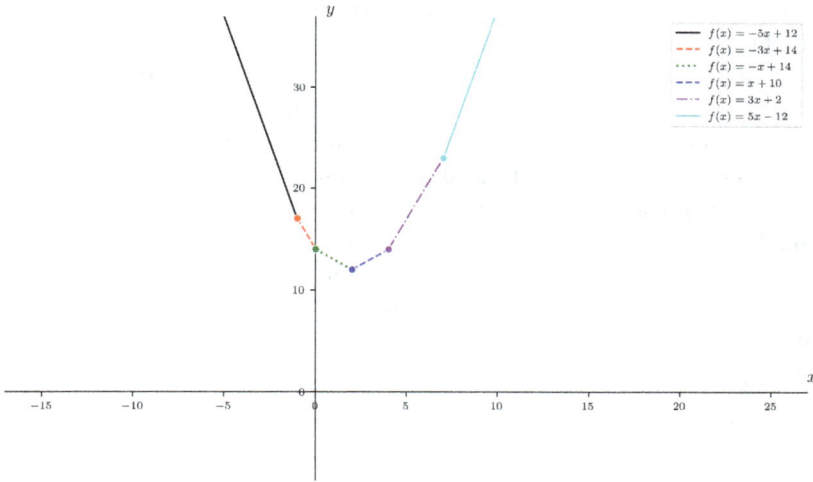

Figure 6.1. The sum of the absolute residuals.

of values in the series, the entire interval between the two middle values has a slope of 0 for $h(x)$, and all the points in it constitute the set of minimum points of the loss function $h(x)$.

Assume for simplicity that n is an even number. As before, the minimal loss is $h\left(med(b)\right) = \sum_{i=1}^{n} |b_i - med(b)|$. It can be seen that for half of the values, namely those that are greater than or equal to the median, the sign does not change because a non-negative number is obtained, while for the other half of the values, namely those that are less than or equal to the median, the sign changes from non-positive to non-negative. Therefore, we can conclude that the value of the loss function at the median is the difference between the sum of the upper and lower halves. That is, if the series is with $b_1 \le b_2 \le \cdots \le b_n$, which can be assumed without loss of generality, then

$$
h\left(med\left(b\right)\right) = \begin{cases} \displaystyle\sum_{i=\frac{n+1}{2}}^{n} b_i - \sum_{i=1}^{\frac{n-1}{2}} b_i, & \text{if } n \text{ is odd,} \\[2em] \displaystyle\sum_{i=\frac{n}{2}+1}^{n} b_i - \sum_{i=1}^{\frac{n}{2}} b_i, & \text{if } n \text{ is even.} \end{cases}
$$

Going back to the general case, our objective here is to minimize the sum of the absolute values of the residuals, $b_i - \Sigma_{j=1}^{m} A_{ij} x_j$, $1 \leq i \leq n$. This is an unconstrained optimization problem which the objective function is not linear.[a] In fact, at points x, where $h(x) = 0$, $h(x)$ is not a differentiable function. Yet, we show next that it can be formulated as a linear program, namely a constrained optimization problem where the objective function is nevertheless linear and where the same is the case with the equality and inequality constraints.

Consider (6.2). Let d_i' and d_i'', $1 \leq i \leq n$, be $2n$ non-negative numbers such that for a given $x \in R^m$, they obey

$$b_i = \Sigma_{j=1}^{m} A_{ij} x_j + d_i' - d_i'', \quad 1 \leq i \leq n.$$

Clearly, if a pair of vectors $d' \in R^n$ and $d'' \in R^n$ satisfy these n equations, then the same is the case with $d' + v$ and $d'' + v$ for any $v \in R^n$. However, consider the following linear program:

$$\min_{x \in R^m, d', d'' \in R^n} \sum_{i=1}^{n} d_i' + \sum_{i=1}^{n} d_i'', \tag{6.4}$$

$$\text{s.t.} \quad Ax + Id' - Id'' = b,$$

$$d', d'' \geq \underline{0}.$$

Here, we have $m + 2n$ decision variables, n equality constraints, and $2n$ inequality constraints. It is easy to see that in the optimal solution $d_i' d_i'' = 0$, $1 \leq i \leq n$. In the case where $d_i' > 0$ (respectively, $d_i'' > 0$), the residual is negative (respectively, positive). Finally, the optimal solution leads to the minimum sum of the absolute values of the residuals.

6.2.2 *Minimizing the maximum absolute residual*

The second objective we shall consider is $\min_{x \in R^n} h(x)$, where

$$h(x) = \max_{i=1}^{n} \left| b_i - \sum_{j=1}^{m} A_{ij} x_j \right|. \tag{6.5}$$

This is known as the ℓ_∞-norm regression, where the goal is to minimize the maximum absolute value among the residuals. In the case where $m = 1$

---

[a]Note that $|x|$ is not a linear function.

and $A_{i1} = 1$, $1 \leq i \leq n$,

$$x^* = \arg\min_{x \in R} \max_{1 \leq i \leq n} |b_i - x|$$

is the mid-range point of the entries b_i, $1 \leq i \leq n$. In other words, it is the average between the smallest and the largest entries among b_i, $1 \leq i \leq n$. Also, assuming $b_1 \leq b_n \leq \cdots \leq b_n$, we get that $x^* = (b_1 + b_n)/2$ and

$$\max_{i=1}^{n} |b_i - x^*| = b_n - x^* = x^* - b_1 = \frac{b_n - b_1}{2}.$$

Example (cont.). In this case, $b_1 = -1$ and $b_5 = 7$. Hence, $x^* = (7 + (-1))/2 = 3$, and $\max_{i=1}^{5} |b_i - x^*| = 3 - (-1) = 7 - 3 = (7 - (-1))/2 = 4$. See Figure 6.2.

Next, we turn to (6.5) and show how it can be formulated as a linear program. Note that, as one now, it is presented as an unconstrained optimization problem, yet with a non-differentiable objective function. For a given $x \in R^m$, let d be any (scalar) that bounds the absolute value of the distances of the b_i from the "fitted" values $(Ax)_i$, $1 \leq i \leq n$. Then, we have the following vector inequality:

$$-d\underline{1} \leq Ax - b \leq d\underline{1},$$

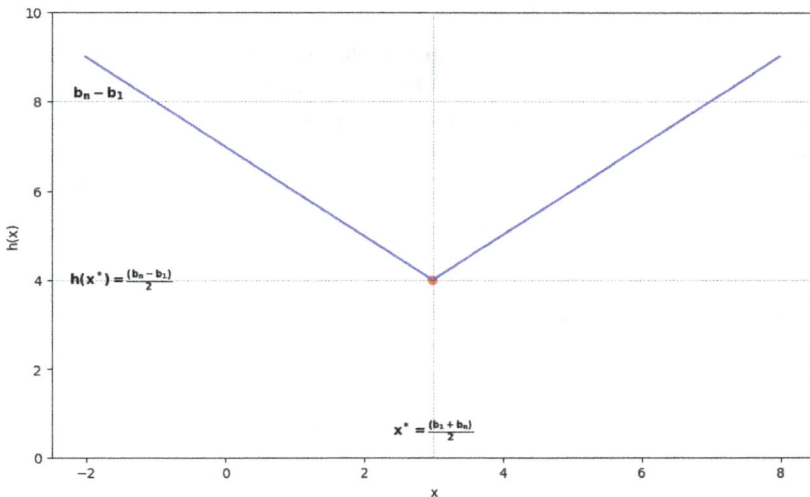

Figure 6.2. The maximum sum of the absolute residual.

where $\underline{1}$ is a vector of ones. This implies that our problem can be formulated as the following linear program:

$$\min_{d \in R, x \in R^m} d, \tag{6.6}$$

$$\text{s.t.} \ -(Ax)_i + d \geq -b_i, \quad 1 \leq i \leq n,$$

$$(Ax)_i + d \geq b_i, \quad 1 \leq i \leq n,$$

which is a linear program with $m + 1$ decision variables and $2n$ inequality constraints. Note that $d \geq 0$ for any feasible (d, x) solution to this program.

For more on the two linear programs stated in (6.4) and (6.6), and in particular, on some of the properties shared by the optimal solutions, see [21] or [15], Chapter 10.

6.3   Example: The Transportation Problem

Suppose that there exist m warehouses and n shops. For some commodity, an amount of $a_i > 0$ is stored in warehouse i, $1 \leq i \leq m$, and the demand in shop j equals $b_j > 0$, $1 \leq j \leq n$. Assume balance, namely $\Sigma_{i=1}^m a_i = \Sigma_{j=1}^n b_j$. Suppose that it costs c_{ij} to ship one unit of the commodity from warehouse i to shop j, $1 \leq i \leq m$, $1 \leq j \leq n$. The goal is to find a shipping scheme that minimizes the total shipping costs, while meeting all the supply and demand requirements. Let the decision variables be x_{ij}, denoting how much to ship from warehouse i to shop j, $1 \leq i \leq m$, $1 \leq j \leq n$. Note that the balance assumption implies that when all supply and demand requirements are met, this is done with equality. In other words, no warehouse ends up with a shortage, and no shop with a surplus. This optimization problem is given by

$$\min_{x_{ij}, 1 \leq i \leq m, 1 \leq j \leq n} \sum_{i=1}^m \sum_{j=1}^n c_{ij} x_{ij}, \tag{6.7}$$

$$\text{s.t.} \ \sum_{j=1}^n x_{ij} = a_i, \quad 1 \leq i \leq m,$$

$$\sum_{i=1}^m x_{ij} = b_j, \quad 1 \leq j \leq n,$$

$$x_{ij} \geq 0, \quad 1 \leq i \leq m, \quad 1 \leq j \leq n.$$

This is known as the *transportation problem* (TB). Observer that a TP is always feasible. For example, the solution

$$x_{ij} = \frac{a_i b_j}{\sum_{k=1}^{m} a_k}, \quad 1 \leq i \leq m, \quad 1 \leq j \leq n \tag{6.8}$$

meets all the constraints.

There are many applications of the TB. For example, in the case where all right-hand side entries are positive and $\Sigma_{i=1}^{m} a_i = 1$ and $\Sigma_{j=1}^{n} b_j = 1$, both the supply and demand vectors can be treated as distribution functions. Then x_{ij} is the (joint) probability of the (i,j)-th cell, $1 \leq i \leq m$, $1 \leq j \leq n$. Suppose a sampling lottery based on this distribution is performed among individuals that belong to some population. If the selected individual belongs to cell (i,j), a cost of c_{ij} is incurred, $1 \leq i \leq m$, $1 \leq j \leq n$. In this case, the goal is to determine the joint distribution with the minimum expected cost among all those whose marginal distributions are as stated by the two probability vectors. To be more concrete, suppose that n points lie on a line. Point i is located d_i meters away from the origin, $1 \leq i \leq n$. There are two individuals, Alice and Bob. Alice is placed at point i with probability a_i, $1 \leq i \leq n$. Define b_i likewise for Bob, $1 \leq i \leq n$. Assume that a cost of $c_{ij} = |d_i - d_j|$ is incurred in the case where Alice is located at point i and Bob at point j, $1 \leq i,j \leq n$. What is at your discretion is deciding on a one-shot two-dimensional lottery for deciding where Alice and Bob should be located at. Your goal is to minimize the expected cost, but we must keep the marginal probabilities as stated above for both Alice and Bob. The resulting LP is therefore

$$\min_{x_{ij}, 1 \leq i \leq n, 1 \leq j \leq n} \sum_{i=1}^{n} \sum_{j=1}^{n} |d_i - d_j| x_{ij}, \tag{6.9}$$

$$\text{s.t.} \quad \sum_{j=1}^{n} x_{ij} = a_i, \quad 1 \leq i \leq n,$$

$$\sum_{i=1}^{n} x_{ij} = b_j, \quad 1 \leq j \leq n,$$

$$x_{ij} \geq 0, \quad 1 \leq i \leq n, \quad 1 \leq j \leq n.$$

Note that the feasible solution stated in (6.8) suggests locating Alice and Bob independently, each with his/her preassigned distribution.

The transportation problem is revisited later in Section 7.2.1. In particular, it is argued there that if a_i, $1 \le i \le m$ and b_j, $1 \le j \le n$ are integers, then (at least) one optimal solution x_{ij}, $1 \le i \le m$, $1 \le j \le n$, is an integer.

Chapter 7

The Simplex Method

7.1 Preliminaries

Definition

Definition 7.1. A linear program is called *standard* if it has the following shape:

$$\min_{x \in R^n} \ c^T x,$$

$$\text{s.t.} \ \ Ax = b,$$

$$x \geq \underline{0}.$$

For some matrix $A \in R^{m \times n}$ with $n \geq m$, it is called the *constraint matrix*; for some vector $b \in R^m$, it is called the *right-hand side* vector; for some vector $c \in R^n$, it is called the *cost vector*; and for some function $c^T x$, it is called the *objective function*. Finally, the variables $x = (x_1, x_2, \ldots, x_n)$ are the *decision variables*.

From now on, we consider only standard linear programs, and hence whenever we say a linear program, we mean a standard one.

Definition

Definition 7.2.

1. Any vector x in R^n is called a *solution*. If $Ax = b$ and $x \geq 0$, then it is called a *feasible solution*. Finally, x^* is called an *optimal solution* if it is feasible and if $c^T x^* \leq c^T x$ for any feasible solution x.
2. A linear program is called *feasible* if it has a feasible solution. Otherwise, it is called *infeasible*.
3. A linear program is called *unbounded* if it is feasible and if an optimal solution does not exist.

Remark. Logic dictates that a fourth option is missing from the above set of definitions, namely that the LP is feasible and bounded but does not have an optimal solution. We will show later that if a linear program is feasible and has no optimal solution, then one can find a feasible solution whose objective function value is as low as one wishes. In other words, the program is *unbounded*. Note also that a necessary condition for unboundedness is that the feasible set itself is unbounded. Recall that a set $X \subseteq R^n$ is said to be bounded if there exists a finite (positive) constant b such that $|x_i| \leq b$, $1 \leq i \leq n$, for any $x \in X$. Otherwise, it is said to be unbounded.

Theorem

Theorem 7.1. *For a standard linear program, the following conditions must be satisfied:*

1. *The set of feasible solutions is convex.*
2. *The set of optimal solutions is convex.*
3. *The set of right-hand side vectors for which x is feasible (respectively, optimal) is convex.*
4. *The set of cost vectors for which x is feasible (respectively, optimal) is convex.*

Exercise 7.1

Prove Theorem 7.1.

The main purpose of this chapter is to introduce the simplex method for solving standard LPs. To this end, we impose some technical assumptions.

Note that they need to be invoked only in the cases where the simplex method is applied.

Assumption 1. The matrix A is a full-rank matrix, namely $rank(A) = m$. This assumption is equivalent to saying that at least one $m \times m$ submatrix of A is invertible.

Remark. Assumption 1 is also equivalent to saying that the m rows of A are linearly independent. Otherwise, there would have been one row, without loss of generality, row m, such that for some (not necessarily unique) $\alpha_1, \alpha_2, \ldots, \alpha_{m-1}$,

$$A_m = \sum_{i=1}^{m-1} \alpha_i A_i, \qquad (7.1)$$

where A_i is the ith row of A, $1 \leq i \leq m$. Then, there are two possibilities. The first is that $\Sigma_{i=1}^{m-1} \alpha_i b_i = b_m$ for at least one of the series α_i, $1 \leq i \leq m-1$, which obey (7.1), in which case, the mth row is redundant: Any $x \in R^n$ that meets the first $m-1$ equality constraints also meets the mth. The second is that $\Sigma_{i=1}^{m-1} \alpha_i b_i \neq b_m$ for at least one of the series α_i, $1 \leq i \leq m-1$, which obey (7.1). In this case the program is not feasible.

Definition

Definition 7.3. Let $B \in R^{m \times m}$ be a (square) submatrix of A. Without loss of generality, assume that B represents the first m columns of A. It is called a *basis* if B^{-1} exists. The variables associated with the basis B are called the *basic* variables, and they are denoted by $x_B \in R^m$. The rest of the matrix A is denoted by N, and the associated variables are called the *non-basic* variables. Denote them by $x_N \in R^{n-m}$. Note that $x = (x_B, x_N)$. The *basic solution* associated with the basis B is $x_N = \underline{0}$ and[a] $x_B = B^{-1}b$. The basis is called *feasible* if[b] $B^{-1}b \geq \underline{0}$. Otherwise, it is called an *infeasible basis*.

---

[a]This selection for x_N and x_B is in order to guarantee that $Ax = b$.
[b]Then $x \geq 0$.

Note that the set of equality constraints can be now written as $Bx_b + Nx_N = b$. The following theorem is now immediate.

Theorem

Theorem 7.2. *If B is a basis and x is a feasible solution, then*

- $x_B = B^{-1}(b - Nx_N)$,
- $c^T x = c_B^T B^{-1} b + (c_N^T - c_B^T B^{-1} N) x_N$,
- $B^{-1} N^j$ *is the vector presentation of N^j as a linear combination of the vectors in the basis[a] B.*

---

[a] N^j is the column of N associated with the variable x_j.

Note that from the second bullet point, we learn that the value of the objective function at the basic solution, which corresponds to the basis B, equals $c_B^T B^{-1} b$.

Remark. Note that once a basis is fixed, the LP is equivalent to

$$c_B^T B^{-1} b + \min_{x_N \in R^{n-m}} (c_N^T - c_B^T B^{-1} N) x_N,$$

$$\text{s.t.} \quad B^{-1} N x_N \leq B^{-1} b,$$

$$x_N \geq \underline{0}.$$

Indeed, the objective function is stated only in terms of x_N, the non-basic variables. Also, the first set of inequalities is in fact the requirement that $x_B \geq \underline{0}$, and the second set is the requirement that $x_N \geq \underline{0}$, making the two sets of inequalities equivalent to the requirement that $x \geq \underline{0}$. Note that now we have only $n - m$ decision variables, while in the original version there were n such variables.

Theorem

Theorem 7.3. *The number of feasible bases is bounded by $n!/(m!(n-m)!)$.*

Exercise 7.2

Prove Theorem 7.3.

7.2   On Bases and Basic Solutions

The notions of a basis and a basic solution were defined in the previous section. The simplex method, to be defined in the next section, will show that, one way or another, everything there is to know about a linear program can be derived from examining only its basic solutions. The following theorem is no exception.

Theorem

Theorem 7.4. *A linear program is feasible if and only if it has a basic feasible solution.*

Proof. The "if" part is of course trivial. What we next show is how to construct a basic feasible solution from some other feasible solution. Specifically, let $x \neq \underline{0}$ be a feasible solution, and without loss of generality, assume that its first few entries are positive and the rest are zero. In particular, let $p \geq 1$ be such that $x_i > 0$, $i = 1, 2, \ldots, p$, and $x_i = 0$, $i = p + 1, \ldots, n$. If the first p columns of A are linearly independent, we are done. Note that if $p \leq m$, then by Assumption 1, some other columns of A can be added to the first p in order to form a basis. Otherwise, namely where $p > m$, let α_i, $1 \leq i \leq p$, be a set of scalars, not all of which are zeros, such that the corresponding linear combination of the first p columns of A is the zero vector in R^n, i.e., $\Sigma_{j=1}^p \alpha_j A^j = \underline{0}$. Note that we can assume one of these coefficients is negative; otherwise, they can all be replaced by their negative values and serve the same purpose. Let $\alpha \in R^n$ be the vector of these p coefficients appended with zeros (to make it a vector in R^n). Clearly, for any scalar ϵ,

$$A(x + \epsilon \alpha) = b.$$

Now, let

$$\epsilon^* = \min_{i, \alpha_i < 0} -\frac{x_i}{\alpha_i}. \tag{7.2}$$

It is easy to check that not only $A(x + \epsilon^* \alpha) = b$, but also $x + \epsilon^* \alpha \geq \underline{0}$, making $x + \epsilon^* \alpha$ a feasible solution. Moreover, this solution has (at least) one more entry with a value of zero than x possesses: this is the entry where the minimization in (7.2) is attained. Now, remove this column from the set of p columns and repeat this procedure for the remaining $p - 1$ columns. The argument can repeat itself from this new solution until we get a solution in which all the columns corresponding to the positive entries are linearly independent. □

The algorithmic conclusion from Theorem 7.4 is clear. Specifically, if one likes to check whether a linear program is feasible or not, all one has to do is to check all basic solutions for feasibility. If none of the solutions are feasible, then the program itself is not feasible. This is a major step forward, since it means that there exists a *finite* algorithm for checking the feasibility of a linear program. This is the case as the number of bases is finite. In particular, this can be executed by a computer. This is apparently the single most important result in linear programming: A continuous optimization problem with a potentially indefinite number of checks that are needed in order to determine feasibility is, in fact, a finite one. In the next section, this property will be demonstrated, and a similar result with respect to determine whether a feasible LP has an optimal solution or is unbounded will be proven. In particular, we will show that in the former case, there is always a basic solution that is optimal.

For a convex set, a point is said to be an *extreme point* if it cannot be expressed as a linear combination of two different points that belong to this set. Since the set of feasible solutions is convex, we can say that a feasible solution is an extreme point of the feasible set if it cannot be expressed as a linear combination of two different feasible solutions. The following theorem characterizes the set of extreme points among the feasible solutions.

Theorem

Theorem 7.5. *A feasible solution is an extreme point of the set of feasible solutions if and only if it is a basic solution.*

Proof. Consider the non-basic solution as defined in the proof of Theorem 7.4, in which we saw that $x + \epsilon\alpha$ for any $\epsilon > 0$ small enough is also a feasible solution. The same can be said of $x - \epsilon\alpha$. Of course, x is the simple average of $x + \epsilon\alpha$ and $x - \epsilon\alpha$. Hence, we can conclude that any non-basic solution is not an extreme point. For the converse, assume that x is a basic solution. We next argue that it is an extreme point of the feasible set. In particular, it cannot be expressed as a linear combination of two different feasible solutions. To prove this by the way of contradiction, assume that x is a basic solution. Then, $x_N = \underline{0}$ and $x_B = B^{-1}b$ for some basis B. Let

x_1 and x_2 be two different feasible solutions such that $x = \alpha x_1 + (1 - \alpha)x_2$ for some $0 < \alpha < 1$. Clearly, the non-basic parts of both x_1 and x_2 need to be zero, as they both are non-negative and their average is zero. This forces the basic parts, both being equal to $B^{-1}b$. Hence, $x_2 = x_1$, which is a contradiction. $\square$

Example 1. Consider the following canonical LP:

$$\max_{x_1, x_2} \; x_1 + x_2,$$

$$\text{s.t.} \; x_1 + 3x_2 \le 30,$$

$$2x_1 + x_2 \le 20,$$

$$x_1, x_2 \ge 0.$$

As depicted in Figure 7.1.

It is possible to see from Figure 7.1 that the feasible set has four extreme points: $x^1 = (0, 0)$, $x^2 = (0, 10)$, $x^3 = (6, 8)$, and $x^4 = (10, 0)$. In order to standardize the LP as stated in Theorem 6.1, we introduce two *slack variables*, namely $s_1 \ge 0$ for the first constraint and $s_2 \ge 0$ for the second:

$$x_1 + 3x_2 + s_1 = 30,$$

and

$$2x_1 + x_2 + s_2 = 20.$$

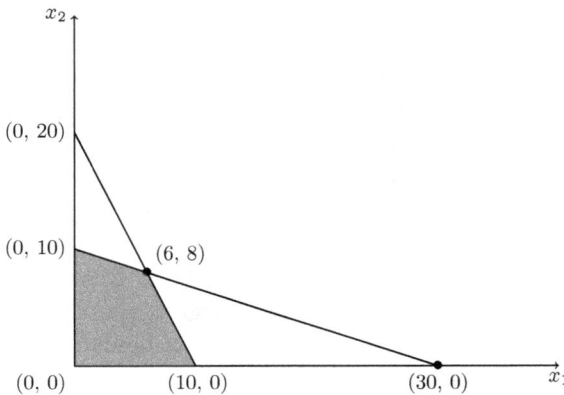

Figure 7.1. A two-dimensional graphic representation.

The corresponding basic variables for the above four extreme points are (s_1, s_2), (x_2, s_2), (x_1, x_2) and (x_1, s_1), respectively. Note that there are two more bases, namely those with (x_1, s_2) and (x_2, s_1) as basic variables, but both correspond to two infeasible solutions, $(30, 0)$ and $(0, 20)$.

In fact, we can say more:

Theorem

Theorem 7.6. *If an optimal solution exists, then a basic optimal solution exists.*

Proof. Let us go back to the non-basic solution that was introduced in the proof of Theorem 7.4, but now assume that it is an optimal solution. Recall the definition of α there, but now note that $-\alpha$ can serve the same purpose. Thus, we can choose either one of them. Hence, we can assume α if $c^T \alpha \le 0$; otherwise, if $c^T \alpha > 0$, replace α with $-\alpha$. This means that $x + \epsilon^* \alpha$ defined above has a value that is not higher than the original objective function value. In particular, $x + \epsilon^* \alpha$ is optimal. If it is a basic solution, we are done. If not, we repeat the procedure stated in the proof of Theorem 7.4 (initiating with $x + \epsilon^* \alpha$, etc.) until we reach a basic solution that is optimal. $\qquad\square$

Let us go back to Example 1. There are four feasible basic solutions, $x^1, \ldots, x^4$. Inserting each one of them in the objective function (note that the coefficients of s_1 and s_2 are zero) yields the values 0, 10, 14, and 10, respectively. Based on Theorem 7.6, we conclude that $x^3 = (6, 8)$ is the optimal solution. Of course, had the objective been to minimize the same objective function, the optimal solution would have been $x^1 = (0, 0)$.

Theorem 7.6 leads to a similar conclusion as the one derived from Theorem 7.4: In the search for an optimal solution, we can limit the search only to basic solutions, which again makes it a finite search. Indeed, one needs to compute the objective values of all feasible basic solutions and take the one (or ones) with the smallest value. But here we need to be careful. If an optimal solution exists, we are done. But what if one doesn't exist? In such a case, there are two more logical possibilities. The first is that the problem is bounded, but the least lower bound cannot be achieved by any feasible solution; namely, we can get as close as we wish to this bound with some feasible solution, without actually obtaining the least lower bound. The second possibility is that the solution is unbounded in the sense that any smaller objective value can be reached by some feasible solution. In the next section, we show that the former possibility is never the case. This cannot be said of the latter. Moreover, we will show that once we have obtained the basic feasible solution with the smallest objective value among all basic feasible solutions, there exists a verifiable condition (to be specified there) that tells whether the solution is bounded (and hence whether an optimal solution exists) or whether it is unbounded.

7.2.1 *Example: the transportation problem (continued)*

The transportation problem was introduced in Section 6.3. It has mn variables and $m + n$ equality constraints. Since $m + n < mn$, the rank of the constraint matrix is bounded by $m + n$. We next argue that its rank equals $m + n - 1$. First, observe that if one subtracts the sum of the rows of the supply constraints from the corresponding sum of the demand constraints, one gets the mn-dimensional zero vector. This is the case since each of the x_{ij} variables, $1 \leq i \leq m$, $1 \leq j \leq n$, appears exactly once in the former set of constraints and exactly once in the latter set. Hence, the $m + n$ rows are not linearly independent, making the rank of the constraint matrix equal to at most $m + n - 1$. In particular, due to symmetry, any one of the $m + n$ equality constraints can be removed without enlarging the set of feasible solutions.

It remains to show that the rank of the constraint matrix is exactly $m + n - 1$. This is done with the help of the lemma below, but first, we need to define some concepts from graph theory. Consider an undirected graph with $m + n$ nodes, where m nodes are associated with the m warehouses and n nodes are associated with the n shops. Moreover, the network has mn arcs, where each arc begins in one of the warehouse nodes and ends in one of the shop nodes. A spanning tree is a set of arcs connecting all nodes such that if one moves along these arcs, a node can be revisited only if one reverses one's moves. It is well known that a tree comes with a number of arcs that is one less than the number of nodes, in our case $m + n - 1$ arcs. Note that not all sets of $m + n - 1$ arcs form a spanning tree. Moreover, when they are not, two things happen: (1) connectivity is lost, and (2) at least one circuit is formed.[a] Note that any node that lies along the circuit can be revisited without backtracking when one moves along the arcs.

Lemma

Lemma 7.1. *In the graph of the transportation problem and its LP formulation, but with one row removed, there is a one-to-one correspondence between spanning trees and bases.*[a]

---

[a]For what is required here, only one direction is needed, namely showing that any given basis forms a spanning tree.

Proof. Consider a spanning tree in the graph associated with a transportation problem with one constraint being removed. There is an arc in it that crosses the node corresponding to the removed constraint.[b] Its column in the current constraint matrix is a unit vector. After a number of elementary row and column operations, it is possible to make this variable the leftmost one, and the row corresponding to the node at the other edge of that arc, the upper row. Then, look for the next arc that crosses this node (pick one randomly in case there are few). This is the second variable from the left, and the constraint of the node at its other edge is the second constraint. Note that 1 appears on the diagonal, followed by a column full of zeros. Repeat this procedure until you get an upper triangular matrix whose diagonal entries are all equal to one. This is an invertible matrix. In this way, we have shown that the constraint matrix is of rank $m + n - 1$. Thus, any basis corresponds to $m + n - 1$ variables. For the converse, take any set of $m + n - 1$ variables and, by way of contradiction, assume that they do not form a spanning tree. Hence, they form a closed circuit. Due to the nature of the graph, this circuit comes with an even number of arcs. Decide on some direction along this circuit, make one arc with a positive sign, and the other one with a negative sign, and so on. When you sum up the columns of these variables, minding these signs, you get the zero vector. This shows that the matrix we deal with comes with columns that are not linearly independent, and hence it is not a basis matrix.   □

---

[a]A circuit is a set of nodes connected by arcs where at least one node is repeated.

[b]Recall that each row in the constraint matrix corresponds to a node and each column to an arc.

From the lemma, it follows immediately that the rank of the constraint matrix is $m + n - 1$:

> **Theorem**
>
> **Theorem 7.7.** *The rank of the constraint matrix of a transportation problem equals the number of warehouses plus the number of shops minus one.*

Using the above proof, we can infer one more thing: the set of equations $Bx_B = b$ can be solved via backwards substitutions. This is the case since B is the upper diagonal matrix with all diagonal entries equal to 1. Hence, no divisions are performed when the equations are solved. This implies that if all the entries in b are integers, then all entries in x_B receive integral values. From this, we gather that if the nature of the problem requires all variables, in particular x_B, to be integers (for example, all variables refer to a number of employees or of tracks), then the LP can be solved as is, ensuring that the optimal solution is restricted to integer values and nothing further is required. Otherwise, adding the constraint that x is restricted to integer values (making the problem an integer program (IP) rather than an LP) would make the problem we face a discrete one (as opposed to a continuous one, as assumed throughout this text). In general, IPs are harder to deal with than LP.

7.3   The Simplex Method

In this section, we introduce the *simplex method*, which receives a standard linear program P and a feasible basis B as input. Its output is an optimal solution (which turns out to be basic) or an indication that P is unbounded. Later on, we will append a preliminary step that finds an initial feasible basis or indicates that the program is infeasible. It was shown above in Theorem 7.4 that P is feasible if and only if it has a feasible basic solution, and hence we have covered all logically possible cases.

7.3.1   *The reduced cost*

The method works as follows. Suppose that you have a feasible basis B with the corresponding solution $(x_B, x_N) = (B^{-1}b, 0)$. Look at the second item in Theorem 7.2. It states the objective function only in terms of the non-basic variables. Note from the first item there that the constraints imply that once the set of non-basic variables x_N are assigned to some values, there is no more freedom in the selection of the basic variables. In fact, x_B is an affine function of x_N. Look also at the vector of *reduced costs* $\bar{c}_N \equiv c_N^T - c_B^T B^{-1} N$. Each of its variables tells how much the objective function changes when the corresponding non-basic variable is increased by one unit. Let $j \in N$ and consider its reduced cost $(\bar{c}_N)_j \equiv (c_N^T - c_B^T B^{-1} N)_j$. If it is non-negative, one cannot improve the value of the objective function by increasing this variable (while keeping all other non-basic variables fixed), which currently has the value of zero. Of course, one would like to reduce its value, but since its value is currently zero, this leads to the violation of the non-negativity constraint. We thus conclude that if $\bar{c}_N \geq \underline{0}$ (a vector inequality), then the basic solution is optimal.

7.3.2   *Improving a basic solution*

Suppose now that $(\bar{c}_N)_j < 0$ for some $j \in N$. Hence, one would like to increase $(x_N)_j$ from zero in order to improve the value of the objective function. Moreover, one would like to make this increase as large as possible. Thus, the question is: What is the maximal possible change in $(x_N)_j$ that maintains the feasibility of the resulting solution? The answer to this question lies with the corresponding changes to the variables of x_B. Specifically, since $x_B = B^{-1}(b - N x_N)$, any change to x_N implies a change to x_B (as long as one maintains its feasibility). Recall that all other non-basic variables remain unaffected. Then,

$$(x_B)_i = (B^{-1} b)_i - (B^{-1} N)_{ij} (x_N)_j, \quad i \in B, \qquad (7.3)$$

namely under these circumstances, $(x_B)_i$ is an affine function of $(x_N)_j$. See Figure 7.2. If $(B^{-1} N)_{ij} \leq 0$, then $(x_B)_i$ is an increasing function of $(x_N)_j$. Hence, in this case, $(x_B)_i$ stays non-negative no matter how large $(x_N)_j$ is. We conclude that if for some $j \in N$, $(\bar{c}_N)_j < 0$, and for all $i \in B$, $(B^{-1} N)_{ij} \leq 0$, then the program is unbounded.

On the other hand, if $(B^{-1} N)_{ij} > 0$, then increasing $(x_N)_j$ too much leads to the violation of the non-negativity constraint, $(x_B)_i \geq 0$. Actually, as far as $(x_B)_i$ is concerned, the largest value that $(x_N)_j$ can get is zero

$(x_B)_i$, namely[c]

$$\frac{(B^{-1}b)_i}{(B^{-1}N)_{ij}}.$$

Since we like to maintain the non-negativity of all variables x_B, the largest value that $(x_N)_j$ can take is

$$\min_{i\in B,(B^{-1}N)_{ij}>0} \frac{(B^{-1}b)_i}{(B^{-1}N)_{ij}}. \tag{7.4}$$

We denote the index where this minimization takes place by i^*. In other words,

$$i^* \equiv \arg \min_{i\in B,(B^{-1}N)_{ij}>0} \frac{(B^{-1}b)_i}{(B^{-1}N)_{ij}}. \tag{7.5}$$

Ties are possible here, and this is an issue we will comment on shortly. The selection of i^* as done in (7.5) is called the *ratio test*. Unless $(B^{-1}b)_{i^*} = 0$, we have obtained a new solution. This is a serious issue since if a basic variable receives a value of zero and is the next variable to be removed from the basis, then no improvement in the objective function value takes place. Therefore, in order to guarantee detecting a strictly better feasible solution, we impose the following assumption.

Assumption 2. If B is a feasible basis, then $x_B = B^{-1}b >> 0$. This is a technical assumption called the *non-degeneracy* assumption. It holds in many practical problems but is not applicable for problems with a special structure, one of which, for example, is the transportation problem defined in Section 6.3.

In the new solution, a previously zero variable becomes positive, and a previously positive variable becomes zero. All other non-basic variables

---

[c]As mentioned, there is a one-to-one correspondence between columns in B and basic variables. This induces a similar correspondence between rows in B and basic variables. For example, if x_7 corresponds to the third column of B, then the same is the case with the third row in B.

maintain their zero value, and all other basic values become (see (7.3))

$$(x_B)_i = (B^{-1}b)_i - (B^{-1}N)_{ij}\frac{(B^{-1}b)_{i^*}}{(B^{-1}N)_{i^*j}}, \quad i \in B.$$

We will argue later that Assumption 2 also implies that i^*, as defined in (7.5), is unique; namely there are no ties in the minimization leading to the decision on the removed variable. See Figure 7.2. In particular, there is exactly one basic variable that now attains the value of zero. This leads to the conjecture that the new solution is also a basic solution, but with respect to another basis, one in which a column of the previous basis (that is associated with i^*) has been replaced by another column (that is associated with $(x_N)_j$). This conjecture will be shown shortly to be correct by proving that the new B matrix is invertible.

By making this change in the basic solution, which stems from changing only one non-basic variable, we end up having a new solution with a reduced objective value in comparison with that of the previous solution.

To summarize what we have done so far, note that a non-basic variable that originally had a zero value was increased and now is positive. Also, some or all of the basic variables have changed; some of them have increased, others have decreased, but at least one of them has reached the value of zero. This is an improved solution in terms of having a lower objective function value. Moreover, as we show next, this is a basic solution with the same set of basic variables, except that i^* is replaced with j.

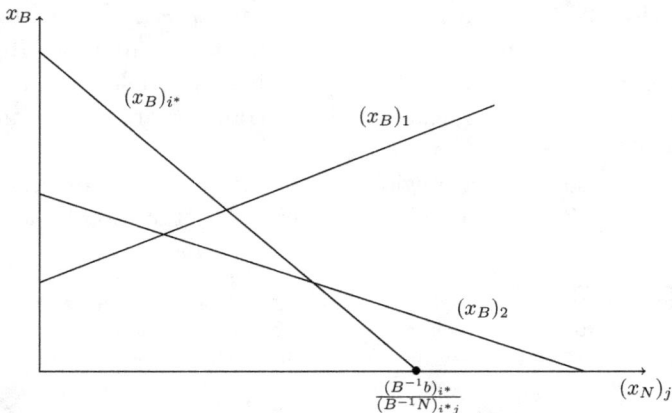

Figure 7.2. Basic variables as functions of a non-basic variable.

Accordingly, the i^* columns in the basic matrix B is replaced with the jth column of N. The procedure can be restarted from this improved solution, and all can now be repeated. Finally, if the ratio test (see (7.4)) is formed over an empty set, the problem is unbounded. However, the fact that the set is not empty does not guarantee the existence of an optimal solution.

Note that having variables x_j and x_{i^*} swap roles as basic and non-basic variables, we get a new set of m variables that constitute a basis.[d] With this new basis, we can repeat all of the above procedure using the new basis. This is what the simplex method is all about. Since at each iteration the value of the objective function improves (by $(c_N^T)_j \frac{(B^{-1}b)_{i^*}}{(B^{-1}N)_{i^*j}}$), and since the number of bases, in particular, those that come with a feasible basic solution, is finite, the procedure eventually halts. That happens either at an optimal solution or by identification that the program is unbounded.

Another issue, which is dealt separately, is how the new matrix inversion is performed. From a theoretical point of view, this is not an issue. Yet, as we will show in Sections 7.5.1 and 7.5.2 below, since only one column was replaced in the previous basic matrix with a new one, some computational effort can be saved.

7.3.3 *The procedure*

We are ready to state the simplex method.
The simplex method

Step 0. Initialization
Let B be a feasible basis.

Step 1. Checking for optimality
Let $\bar{c}_N = c_N^T - c_B^T B^{-1} N$. If $\bar{c}_N \geq 0$, then print "The program is bounded and $x^* = (x_B^*, x_N^*) = (B^{-1}b, 0)$ is an optimal solution." and STOP. Otherwise, let $j \in N$ be an index such that $(\bar{c}_N)_j < 0$. Go to Step 2.

Step 2. Checking for unboundedness
If $(B^{-1}N)_{ij} \leq 0$ for all $i \in B$, then print "The program is unbounded." and STOP. Otherwise, go to Step 3.

---

[d]Later on, in Section 7.5.1, we show that the resulting m columns are linearly independent.

Step 3. Improvement
Let

$$i^* = \arg \min_{i \in B, (B^{-1}N)_{ij} > 0} \frac{(B^{-1}b)_i}{(B^{-1}N)_{ij}}.$$

Let $B \leftarrow B \backslash i^* \cup j$.
Go to Step 1.

Theorem

Theorem 7.8. *The improvement step in the simplex method*

- *generates a new basis.*
- *defines i^* uniquely.*

Proof. The first bullet point is proved later in Section 7.5.1. The key point to observe here is that $(B^{-1}N)_{ij} \neq 0$. Assuming the first bullet point is true, we next deal with the second point. Assume that there is another basic variable, denoted by i^{**}, which ties i^* for the minimization defined in (7.5). Then, when considering the new basis (the one where x_j replaces x_{i^*} as a basic variable), we see that the values of its basic variables change in accordance with (7.3). In particular, due to the tie, $x_{i^{**}} = 0$ too. This violates Assumption 2, the non-degeneracy assumption. □

We have in fact established the following important result.

Theorem

Theorem 7.9. *In each iteration of the simplex method, a new feasible basis is found such that the objective function value of the corresponding basic solution is strictly decreasing. Hence, no basis is repeated. In particular, the method terminates in a finite number of iterations, either with an optimal solution or by showing that the program is unbounded.*

7.3.4   Some remarks

Remark 1. Note that Theorem 7.9 says that (for a feasible program) if an optimal solution exists, the simplex method halts when it reaches such a solution (which as it turns out, is a basic solution). Likewise, if the program is unbounded, then at one of the iterations (not necessarily the first), a basis is found with respect to which the unboundedness criterion is satisfied. This in particular means that, for an unbounded program, the unboundedness criterion is not necessarily satisfied for all bases that the procedure encounters, but for at least one of them (and then unboundedness is declared). This is the case, as otherwise, in the case of an unbounded program, the procedure will run forever. However, the procedure cannot run indefinitely as the number of bases is finite, and at each iteration, the objective value strictly decreases. In particular, the same basis cannot be visited twice.

Remark 2. The above analysis, coupled with the discussion at the end of the previous section, implies that one can do with a more primitive algorithm than the simplex. Specifically, one can compute all basic solutions and check each of them for feasibility. If none of them is feasible, the program is infeasible. Otherwise, for each basic solution, the corresponding value of the objective function is determined, and the one with the minimal value is examined for the optimality criterion. If the optimality criterion is met, the program is bounded and this basic solution is optimal. Otherwise, the program is unbounded.[e,f]

Remark 3. The algorithm does not specify which non-basic variable should be selected to enter the basis in the cases where more than one of them incurs a negative reduced cost. Indeed, any choice will do as far as finite convergence is concerned. A greedy approach would be to select the most negative reduced cost, but an arguably better approach would be to select the first reduced cost encountered, as it would eliminate the computational cost of computing additional reduced costs. This approach has another advantage: In many cases, the columns at the constraint matrix are not computed in advance and they are called for only when their reduced

---

[e]The unboundedness criterion is not necessarily met by all basic solutions (even if the program is unbounded), but it is necessarily met at the basic solution with the minimum objective function value among all basic solutions.

[f]After computing all basic solutions, their objective functions are computed, and they are ordered from the lowest to the highest based on these values. Then, they are checked for feasibility in ascending order. If none is feasible, then the program is infeasible. Otherwise, the optimality criterion is checked on the first encountered feasible basis. In case it is met, the program is bounded, and this is an optimal solution. Otherwise, the program is unbounded.

costs are computed for the first time. This approach is called *column generation*. Finally, since

$$\min_{i \in B, (B^{-1}N)_{ij} > 0} \frac{(B^{-1}b)_i}{(B^{-1}N)_{ij}}$$

is the value of the added variable to the new basis,

$$\arg\min_{j \in N} \left[(\bar{c}_N)_j \times \min_{i \in B, (B^{-1}N)_{ij} > 0} \frac{(B^{-1}b)_i}{(B^{-1}N)_{ij}} \right]$$

leads to the biggest possible improvement in the value of the objective function in the current iteration. This column selection strategy is called the *greatest increment* method.

Remark 4. When one non-basic variable increases and the other non-basic variables remain unchanged, the basic variables are changed accordingly to maintain feasibility. This, in fact, means moving along some direction $d \in R^n$. In particular, $d = (d_B, d_N)$, where $d_B = -B^{-1}N^j$, and d_N consists of zeros except for the entry corresponding to the added variables $(x_N)_j$ is 1. In this direction, the function is reduced by $c^T d$ per unit of change in $(x_N)_j$. The product $c^T d$ equals

$$c_B^T d_B + c_N^T d_N = -c_B^T B^{-1} N^j + (c_N)_j = (\bar{c}_N)_j,$$

which is nothing but the reduced cost of the entering variable (which of course is negative). It is possible to see that this direction is the steepest descent direction among all directions that are based on some convex combination of positive changes at the non-basic variables (and the resulting changes in the basic variables).

Remark 5. We have seen that any choice of a negative reduced cost leads to an improved adjacent basis solution. But the converse is not necessarily true: If one finds an adjacent basic solution, the direction from the previous one to the new one is not necessarily a descending one.

Remark 6. Note that the non-degeneracy assumption was invoked above only once: When we claimed that there is a strict improvement in the objective function in each iteration of the Simplex method. In particular, this assumption is not needed when we check the optimality of a given basic solution or to determine whether the unboundedness condition holds there.

Remark 7. Example 2 in Section 4.5.2 claimes that no duality gap exists in linear programming. We are ready to prove this. Let B be the primal optimal basis that solves (4.12). Consider $\lambda = (c_B^T B^{-1})^T$ as a solution for (4.13). It is easy to see that it is dual feasible:

$$c_B^T B^{-1} A = c_B^T B^{-1}(B, N) = (c_B^T, c_B^T B^{-1} N) \leq (c_B^T, c_N^T).$$

Note that the first part of this inequality is trivial (in fact, it is an equality), while the second part uses the fact that the reduced costs associated with an optimal basis are non-negative. Finally, the dual objective function for this solution equals $b^T (c_B^T B^{-1})^T = c_B^T B^{-1} b$, which coincides with the primal optimal objective value.

7.3.5 A numerical example

Consider the following LP (called the Dakota Furniture example in Winston's text [23]):

$$\max_{x_1, x_2, x_3} \quad 60x_1 + 30x_2 + 20x_3,$$

$$\text{s.t.} \quad 8x_1 + 6x_2 + x_3 \leq 48,$$

$$4x_1 + 2x_2 + 1.5x_3 \leq 20,$$

$$2x_1 + 1.5x_2 + 0.5x_3 \leq 8,$$

$$x_2 \leq 4,$$

$$x_i \geq 0, \quad 1 \leq i \leq 3.$$

The first thing to do is to convert the above LP into a standard LP. Add four slack variables. Denote them by x_4, x_5, x_6, and x_7. The cost vector is then $c^T = (-60, -30, -20, 0, 0, 0, 0)$, the variables are $x = (x_1, x_2, x_3, x_4, x_5, x_6, x_7)$, and the objective is to minimize $c^T x$. The constraint matrix is

$$A = \begin{pmatrix} 8 & 6 & 1 & 1 & 0 & 0 & 0 \\ 4 & 2 & 1.5 & 0 & 1 & 0 & 0 \\ 2 & 1.5 & 0.5 & 0 & 0 & 1 & 0 \\ 0 & 1 & 0 & 0 & 0 & 0 & 1 \end{pmatrix}.$$

The right-hand side vector is $b^T = (48, 20, 8, 4)$. The non-negativity constraints are $x_i \geq 0$, $1 \leq i \leq 7$. Finally, using our notation, $m = 4$ and $n = 7$.

As is always the case where all original constraints are inequalities and the right-hand side entries are positive, the added slack variables can serve as an initial feasible basis. Thus, the simplex commences with $x_B = (x_4, x_5, x_6, x_7)$ and $x_N = (x_1, x_2, x_3)$. The matrix B is the 4×4 identity matrix, and the same can be said of it inverse:

$$B = \begin{pmatrix} 1 & 0 & 0 & 0 \\ 0 & 1 & 0 & 0 \\ 0 & 0 & 1 & 0 \\ 0 & 0 & 0 & 1 \end{pmatrix} \quad \text{and} \quad B^{-1} = \begin{pmatrix} 1 & 0 & 0 & 0 \\ 0 & 1 & 0 & 0 \\ 0 & 0 & 1 & 0 \\ 0 & 0 & 0 & 1 \end{pmatrix}.$$

Observe which column in each of these two matrices corresponds to which of the basic variables. Note that

$$N = \begin{pmatrix} 8 & 6 & 1 \\ 4 & 2 & 1.5 \\ 2 & 1.5 & 0.5 \\ 0 & 1 & 0 \end{pmatrix}.$$

and again note how the columns in N correspond to the non-basic variables.

The first basis. The initial basic solution is of course $x_N = (x_1, x_2, x_3) = (0, 0, 0)$ and $x_B = B^{-1}b$. Then, we can conclude that $x_B = (x_4, x_5, x_6, x_7) = (48, 20, 8, 4)$. Note that $c_B^T = (0, 0, 0, 0)$ and $c_N^T = (-60, -30, -20)$. The first thing is to check for the optimality condition by computing the reduced cost vector and checking its sign. Recall that the reduced cost is $\bar{c}_N = c_N^T - c_B^T B^{-1} N$. In our case, the computations are easy since c_B^T is zero. Hence, $\bar{c}_N = c_N^T = (-60, -30, -20)$. Since the vector of reduced costs includes at least one negative entry, we conclude that the current basic solution is not optimal. We next need to decide which variable, of all those with a negative sign reduced cost, should enter. We have some degrees of freedom here, but due to the convention we imposed, x_1 is the one to enter.

The next thing to do is to check for the unboundedness condition. To this end, we need to multiply B^{-1} with the column corresponding to the added variable x_1:

$$B^{-1} A^{x_1} = \begin{pmatrix} 1 & 0 & 0 & 0 \\ 0 & 1 & 0 & 0 \\ 0 & 0 & 1 & 0 \\ 0 & 0 & 0 & 1 \end{pmatrix} \begin{pmatrix} 8 \\ 4 \\ 2 \\ 0 \end{pmatrix} = \begin{pmatrix} 8 \\ 4 \\ 2 \\ 0 \end{pmatrix}.$$

As at least one of the above entries is positive, the unboundedness condition is not satisfied.

The next question is which variable should be removed from the basis. As there is more than one entry above that is positive, the answer is not immediate. To answer it, we need to perform the following ratio test:

$$\min_{i\in B,(B^{-1}N)_{i1}>0} \frac{(B^{-1}b)_i}{(B^{-1}N)_{i1}} = \min\left\{\frac{48}{8}, \frac{20}{4}, \frac{8}{2}\right\} = 4. \tag{7.6}$$

In fact, the value of 4 we obtained here is of secondary importance. Note that it equals the new value of x_1, the added variable, in the new basic solution. What is more important is where this minimization is achieved, in the ratio corresponding to x_6. Hence, x_6 is removed from the basis.

The second basis. The new set of basic variables is $x_B = (x_4, x_5, x_1, x_7)$. Note the order here: x_1 is placed where x_6 was. This is a compulsory requirement from the Simplex method for efficiency in matrix inversions. See more on this in Section 4. Of course, $x_N = (x_2, x_3, x_6)$ and here, any order is okay. Note, however, that this order has to be kept, at least until a new set of non-basic variables is selected. For this new basis

$$B = \begin{pmatrix} 1 & 0 & 8 & 0 \\ 0 & 1 & 4 & 0 \\ 0 & 0 & 2 & 0 \\ 0 & 0 & 0 & 1 \end{pmatrix} \quad \text{and} \quad B^{-1} = \begin{pmatrix} 1 & 0 & -4 & 0 \\ 0 & 1 & -2 & 0 \\ 0 & 0 & 0.5 & 0 \\ 0 & 0 & 0 & 1 \end{pmatrix}. \tag{7.7}$$

It is worth noting that if one of the columns of a matrix is a unit vector where the 1 appears on the diagonal, the same is the case with the corresponding column in its inverse. Of course, we have three columns with this property:

$$N = \begin{pmatrix} 6 & 1 & 0 \\ 2 & 1.5 & 0 \\ 1.5 & 0.5 & 1 \\ 1 & 0 & 0 \end{pmatrix} \quad \text{and} \quad B^{-1}N = \begin{pmatrix} 0 & -1 & -4 \\ -1 & 0.5 & -2 \\ 0.75 & 0.25 & 0.5 \\ 1 & 0 & 0 \end{pmatrix}.$$

Also,

$$\begin{pmatrix} x_4 \\ x_5 \\ x_1 \\ x_7 \end{pmatrix} = B^{-1}b = \begin{pmatrix} 1 & 0 & -4 & 0 \\ 0 & 1 & -2 & 0 \\ 0 & 0 & 0.5 & 0 \\ 0 & 0 & 0 & 1 \end{pmatrix} \begin{pmatrix} 48 \\ 20 \\ 8 \\ 4 \end{pmatrix} = \begin{pmatrix} 16 \\ 4 \\ 4 \\ 4 \end{pmatrix}.$$

Note that the value for x_1, 4, is as predicted in (7.6).

The remaining variables have a current value of zero. Finally, it is possible to see (by inserting the current values of all of the seven variables in the objective function) that currently the value of the objective function is -240 (which improves upon the initial value of 0). Indeed, we get the same value by computing $c_B^T B^{-1} b$. Note that $B^{-1} b$ has already been computed, so not much is left to do (especially as three out of the four variables in the c_B^T are 0).

We next need to check whether the optimality condition is satisfied with this basis. Note that we never checked the feasibility condition as it is always satisfied due to the way the simplex procedure selects the next basis. Thus, we need to compute the reduced cost, which is defined by $\bar{c}_N = c_N^T - c_B^T B^{-1} N$ which is done next. Recall that $B^{-1} N$ has already been computed above. Thus,

$$\bar{c}_N = (-30, -20, 0) - (0, 0, -60, 0) \begin{pmatrix} 0 & -1 & -4 \\ -1 & 0.5 & -2 \\ 0.75 & 0.25 & 0.5 \\ 1 & 0 & 0 \end{pmatrix}$$

$$= (-30, -20, 0) + (45, 15, 30) = (15, -5, 30).$$

The optimality condition is not satisfied. The only candidate to add to the basis is x_3.

We next need to check the unboundedness criterion. This is done by considering $B^{-1} A^{x_3}$, where A^{x_3} is the column of A corresponding to the variable x_3. Note that this vector has already been computed: It is the middle column in $B^{-1} N$ above. As it contains at least one positive entry, the unboundedness condition is not satisfied here.

If x_3 is added to the basis, which variable is removed from it? The column

$$B^{-1} A^{x_3} = \begin{pmatrix} -1 \\ 0.5 \\ 0.25 \\ 0 \end{pmatrix}$$

contains two positive entries, and hence there are two candidates to remove from the basis: x_5 and x_1. Note that the fact that x_1 was just added does not rule out the possibility that it will soon be removed. Next we perform the ratio test:

$$\min_{i \in B, (B^{-1}N)_{i3} > 0} \frac{(B^{-1}b)_i}{(B^{-1}N)_{i3}} = \min\left\{\frac{4}{0.5}, \frac{4}{0.25}\right\} = 8. \tag{7.8}$$

Again, the value of x_3 in the new basic solution is 8. More importantly, the variable to remove from the basis is x_5 as the minimization in (7.8) takes place at x_5.

The third (and final) basis. The third set of basic variables is $x_B = (x_4, x_3, x_1, x_7)$ and again we note the order of the basic variables. Of course, $x_N = (x_2, x_5, x_6)$.

$$B = \begin{pmatrix} 1 & 1 & 8 & 0 \\ 0 & 1.5 & 4 & 0 \\ 0 & 0.5 & 2 & 0 \\ 0 & 0 & 0 & 1 \end{pmatrix} \quad \text{and} \quad B^{-1} = \begin{pmatrix} 1 & 2 & -8 & 0 \\ 0 & 2 & -4 & 0 \\ 0 & -0.5 & 1.5 & 0 \\ 0 & 0 & 0 & 1 \end{pmatrix}. \tag{7.9}$$

The current values of the basic variables are

$$\begin{pmatrix} x_4 \\ x_3 \\ x_1 \\ x_7 \end{pmatrix} = B^{-1}b = \begin{pmatrix} 1 & 2 & -8 & 0 \\ 0 & 2 & -4 & 0 \\ 0 & -0.5 & 1.5 & 0 \\ 0 & 0 & 0 & 1 \end{pmatrix} \begin{pmatrix} 48 \\ 20 \\ 8 \\ 4 \end{pmatrix} = \begin{pmatrix} 24 \\ 8 \\ 2 \\ 4 \end{pmatrix}.$$

Note that the value for x_3, 8, is as predicted in (7.8). By definition, $x_2 = x_5 = x_6 = 0$. As for the objective value, we need to multiply $c_B^T = (0, -20, -60, 0)$ by x_B. This yields -280, which improves upon -240.

It this the optimal value? The answer is based on checking the reduced cost $\bar{c}_N - c_B^T B^{-1}N$ for this basis:

$$(-30, 0, 0) - (0, -20, -60, 0) \begin{pmatrix} 1 & 2 & -8 & 0 \\ 0 & 2 & -4 & 0 \\ 0 & -0.5 & 1.5 & 0 \\ 0 & 0 & 0 & 1 \end{pmatrix} \begin{pmatrix} 6 & 0 & 0 \\ 2 & 1 & 0 \\ 1.5 & 0 & 1 \\ 1 & 0 & 0 \end{pmatrix}$$

$$= (5, 10, 10).$$

All the entrant variables are non-negative, and hence we declare the current basic solution as optimal and -280 as the optimal value. Recall that for the original maximization LP, the optimal value is 280.

7.4 Phase I: Finding an Initial Feasible Basic Solution

Recall that the Simplex method initializes with a feasible basic solution. We next deal with the question of how such a feasible solution is found (whenever the program is indeed feasible) and how it is discovered that the program is not feasible. Interestingly, this is achieved by applying the Simplex itself to another (related) LP.

Assume, without loss of generality, that $b \geq 0$; otherwise, replace any row whose right-hand side is a negative value by its negative. Consider the following linear program that stems from program P:

$$\min_{x \in R^n, w \in R^m} \sum_{i=1}^{m} w_i,$$

$$\text{s.t.} \quad Ax + Iw = b,$$

$$x \geq \underline{0}, \quad w \geq \underline{0}.$$

The set of m new variables $w \in R^m$ are known as *artificial* variables. Note that $x = \underline{0}$ and $w = b$ are a feasible solution for this program. Moreover, it is a basic solution where the variables $w = (w_1, w_2, \ldots, w_m)$ are the basic variables (with I as the corresponding basis) and the variables $x = (x_1, x_2, \ldots, x_n)$ are the non-basic part. Hence, the simplex method stated above is well-defined for this program as we have an initial feasible basis. However, we still need to assume that the no-degeneracy assumption holds for this program as well. Note that a necessary condition for this to hold is that all entries of b are positive. Also, and clearly, this program is bounded zero. Hence, once applied, the Simplex method terminates with an optimal basis. There are two possible options for the value of the optimal objective function:

- It is strictly positive. Hence, P is infeasible. This is the case as any feasible solution to P (if one exists), coupled with all artificial variables having the value of zero, yields a better objective function, namely zero.
- It is zero. Then, all the artificial variables w ought to have a value of zero. Hence, the corresponding x components of this solution are a feasible solution for the original program. It is likely that this is a basic solution, but this is not always the case. However, this is not an issue, since in those cases where it is not a basic solution, one can apply the procedure described in Theorem 7.2, which yields a feasible basic solution with which the Simplex can be initialized for the original program.

We refer to the above analysis as *Phase I* of the Simplex method, and then what follows it, namely when the simplex method is applied to the original program, is called *Phase II*.

7.4.1 A numerical example

Given the following linear program:

$$\max_{x_1, x_2, x_3} \quad 2x_1 + 3x_2 - 5x_3,$$

$$\text{s.t. } x_1 + x_2 + x_3 = 2,$$

$$-2x_1 + 5x_2 - x_3 = -3,$$

$$x_i \geq 0, \quad i = 1, 2, 3.$$

1. Find a feasible solution (if it exists) by applying Phase I of the simplex method;
2. Solve the LP by the Simplex method. Initialize it with the basis with which Phase I terminates.

Solution. The LP for Phase I is

$$\min_{x_1, x_2, x_3, w_1, w_2} \quad w_1 + w_2,$$

$$\text{s.t. } x_1 + x_2 + x_3 + w_1 = 2,$$

$$2x_1 - 5x_2 + x_3 + w_2 = 3,$$

$$x_1, x_2, x_3, w_1, w_2 \geq 0.$$

An initial basis is composed of the artificial variables w_1 and w_2. Their values are 2 and 3, respectively. The three original variables are non-basic, and hence their value is zero. Note that the value of the objective function is 5. The first thing to do is to express the objective function $w_1 + w_2$ in terms of the non-basic variables x_1, x_2, and x_3. To this end, we note that

$$w_1 = 2 - x_1 - x_2 - x_3$$

and

$$w_2 = 3 - 2x_1 + 5x_2 + x_3.$$

Summing up these two equations, we get that the objective function is $5 - 3x_1 + 4x_2 - 2x_3$. The current solution is not optimal, as at least one of the reduced costs is negative. Accordingly, we include x_1 in the set of basic variables, although the other two variables are possible choices. There is no need to check whether the unboundedness condition is satisfied, as we have already argued that the LP of Phase I is never unbounded. Hence, the next question is which variable should be removed from the basis. The ratio test is

$$x_1 = \min \left\{ \frac{2}{1}, \frac{3}{2} \right\}.$$

Hence, $x_1 = 1.5$ and the variable to remove from the basis is w_2. The value of the objective function is $5 - 3 \times 1.5 = 0.5$, which is not zero yet.

The basic variables are x_1 and w_1. Expressing them in terms of x_2, x_3, and w_2 yields

$$x_1 = \frac{1}{2} \left(3 + 5x_2 - x_3 - w_2 \right)$$

and

$$w_1 = 0.5 - 3.5x_2 - 0.5x_3 + 0.5w_2.$$

Next we need to express $w_1 + w_2$ in terms of x_2, x_3, and w_2. We get that $w_1 + w_2$ equals $0.5 - 3.5x_2 - 0.5x_3 + 1.5w_2$. The current solution is not optimal as some of the reduced costs are negative. Thus, x_2 can be added to the basis. Since we do not need to check for unboundedness, we next ask which variable should now be removed from the basis. In the resulting ratio test, there is only one participant, w_1, and hence it is removed from the basis. Therefore, the next basis is composed of x_1 and x_2. Once none of the artificial variables is basic, Phase I terminates.

Next we turn to solve the original LP.[g] Note that x_1 and x_2 are basic variables, while x_3 is not basic. The values for the basic variables were implicitly determined above: $x_2 = 1/7$ and $x_1 = 13/7$. Next, we need to express x_1 and x_2 as functions of x_3:

$$x_1 = \frac{13}{7} - \frac{6}{7}x_3,$$

$$x_2 = \frac{1}{7} - \frac{1}{7}x_3.$$

---

[g]The objective here it to maximize an objective function so that some inequalities are reversed.

Expressing the objective function in terms of x_3, we get

$$\frac{29}{7} - \frac{50}{7}x_3.$$

Hence, this is the optimal solution and we are done. In summary, the optimal solution is $x_1 = 13/7$, $x_2 = 1/7$, and $x_3 = 0$. The corresponding optimal objective value is $29/7$.

7.5 Updating B^{-1}

The Simplex procedure calls for a matrix inversion per iteration. Usually, the inversion of an $m \times m$ matrix needs a number of numerical operations that are proportional to m^3.[h] However, consecutive basic matrices differ only in one column. Next, it will be shown that once the previous B^{-1} is given, the computational effort needed in order to compute the new B^{-1} is proportional to m^2. We show this in two equivalent ways. The first is what is usually referred to as a *pivot step*. The second is the *Sherman–Morrison formula* for rank-one updates. Details on both procedures are given next.

7.5.1 *Pivot steps*

Inversion of matrices is sometimes performed by applying *elementary row operations*. In short, $(B, I) \to (I, B^{-1})$. Apply the same operations on the rows of the matrix N and get $(B, I, N) \to (I, B^{-1}, B^{-1}N)$. Since one is interested in the inverse of the new basic matrix, interchange the columns $i^* \in B$ and $j \in N$ (but apply the same operations as before). This procedure will change the identity matrix: The column that was previously corresponding to i^* is $B^{-1}N^j$ where N^j is the column corresponding to the previously nonbasic variable x_j. But otherwise, the identity matrix stays intact. In order to get the inverse of the new basis, we have to apply the needed row operations in order to make this non-unit column a unit-one column. We then check whether $(B^{-1}N)_{i^*j}$, which appears on the diagonal of the leftmost $m \times m$ matrix, is indeed the pivotal entry in the matrix B^{-1}. In particular, since it has a non-zero value, the operations will lead to the desired unit vector.[i] Then, we perform m elementary row operations, each requiring an effort proportional to m, leading to a computational effort

---

[h]There are more sophisticated algorithms with reduced complexity but they are beyond the scope of this book.

[i]The fact that a pivotal entry has a non-zero value proves that the new set of columns is invertible, and hence it constitutes a basis.

which is proportional to m^2, as promised. For the next iteration, we will need the new $B^{-1}N$, and therefore, we need to apply the same operations along all the rows of the matrix. As our final computational comment, note that if we use the column generation approach, we do not have to compute $B^{-1}N^j$ for all non-basic columns; it is enough to do so until we get a negative reduced cost.

Remark. Recall the concept of Phase I which was introduced in Section 7.4. It is possible to define a Phase-I LP with a single artificial variable rather than m as suggested above. This is done as follows. Select some (not necessarily feasible) basis B. Let $\bar{b} = B^{-1}b$. If $\bar{b} \geq 0$, there is no need for Phase I. Otherwise, let r be the most negative entry in $\bar{b}$. Now introduce an artificial variable x_{n+1} whose column is $-B\underline{1}$, where $\underline{1}$ is a vector of ones. In particular,

$$Ix_B + B^{-1}N - x_{n+1}\underline{1} = \bar{b}.$$

Now, use the rth entry in the column of x_{n+1} as a pivot. Specifically, after a series of row operations, the x_{n+1}-column becomes a unit vector with 1 at the rth row. Note that the selection of r was made in order to preserve the non-negativity of the right-hand side. Finally, x_{n+1} is added to the basis, while x_r is removed from it. Phase I actually commences here. When it terminates, $x_{n=1}$ is a non-basic variable if and only if the original LP is feasible.

Exercise 7.3

Prove the following technical point: Let A be an invertible matrix with one column that is a unit vector.[j] Show that the corresponding column at A^{-1} is also a unit vector.

7.5.2   *Rank-one updates*

We start with the following result which is known as the *Sherman–Morrison formula*.

---

[j]Recall that by a unit vector in a matrix, we mean that the diagonal entry is one and the off-diagonal entries are zeros.

> **Theorem**
>
> **Theorem 7.10.** *Let $A \in R^{n \times n}$ be an invertible matrix and let $z, x \in R^n$. Then, $A + zx^T$ is invertible if and only if $x^T A^{-1} z \neq -1$. Moreover, in this case,*
>
> $$(A + zx^T)^{-1} = A^{-1} - \frac{1}{1 + x^T A^{-1} z} A^{-1} zx^T A^{-1}. \qquad (7.10)$$
>
> *In particular, once A^{-1} is given, computing $(A + zx^T)^{-1}$ is an $O(n^2)$ computational task.*

Proof. Identity (7.10) can be proved by a straightforward matrix multiplication, which is omitted. In the case where $x^T A^{-1} z = -1$, it is trivial that $z \neq \underline{0}$. However, $(A + zx^T) A^{-1} z = \underline{0}$. Since $A^{-1} z \neq \underline{0}$, the previous equation implies that $A + zx^T$ is not invertible.

As for the computational complexity associated with the use of Sherman–Morrison formula, note that when we multiply a number of matrices, the order of multiplication is irrelevant for the end result, i.e., $ABC = (AB)C = A(BC)$. This is not the case in terms of the computational effort. In particular, an efficient way to compute the updating term $A^{-1} zx^T A^{-1}$ is via $(A^{-1} z)(x^T A^{-1})$. Indeed, the task in each parenthesis is of order $O(n^2)$, and the same is the case with the final column-by-row multiplication. $\qquad \square$

The Sherman–Morrison identity basically says that if a matrix is updated by an additive term that is a rank-one matrix (which can always be represented as an outer product between a column vector and a row vector), the inverse of the new matrix can be computed with an effort proportional to $O(n^2)$ (once the inverse of the original matrix is given). This is more efficient than inverting the new matrix from scratch. Note that the correction term in (7.10) is also a rank-one matrix.

The application of the Simplex method is now straightforward. If B is the old basis and if its ith column is updated, then the new basis equals $B + ze_i^T$, where z is the difference between the column corresponding to the entrant variable and the column corresponding to the removed variable, and where e_i is the ith unit vector. It is tempting to think that the special structure of this update (where e_i is a vector that consists of zeros with one exception, that of entry i, which equals 1) will lead to a further reduction in computational complexity, but this is not the case.

Remark. The fact that the new matrix is invertible was established in the previous section. An independent proof, utilizing Theorem 7.10, is given next. Actually, we need to show that the corresponding $x^T A^{-1} z \neq -1$. Note that in the case under consideration $x = e_{i^*}$, $A^{-1} = B^{-1}$, and $z = N^j - B^{i^*}$. First, $(e^{i^*})^T B^{-1} = B^{-1}_{i^*.}$; that is, $(e^{i^*})^T B^{-1}$ is the i^* row of B^{-1}. Second,

$$x^T A^{-1} z = B^{-1}_{i^*.}(N^j - B^{i^*}) = \sum_k B^{-1}_{i^* k}(N_{kj} - B_{ki^*}) = \sum_k B^{-1}_{i^* k} N_{kj} - 1.$$

Since $\Sigma_k B^{-1}_{i^* k} N_{kj} = (B^{-1} N)_{i^* j} \neq 0$ (see (7.5)), the proof is completed.

Example. In the example given in Section 7.3.5, we did not deal with the issue of how consecutive matrix inversions were performed. We next exemplify how to do so by using the rank-one method suggested in this section.

First, note that the original basis, and hence its inverse, equals I. The new basis is the same as the previous one plus ze_3^T, where

$$z = \begin{pmatrix} 8 \\ 4 \\ 2 \\ 0 \end{pmatrix} - \begin{pmatrix} 0 \\ 0 \\ 1 \\ 0 \end{pmatrix} = \begin{pmatrix} 8 \\ 4 \\ 1 \\ 0 \end{pmatrix},$$

and where e_3 is the third unit vector. Thus, the inverse equals

$$I - \frac{1}{1 + e_3^T I z} I z e_3^T I = I - \frac{1}{2} z e_3^T$$

$$I - \frac{1}{2} \begin{pmatrix} 8 \\ 4 \\ 1 \\ 0 \end{pmatrix} (0,0,1,0) = I - \begin{pmatrix} 0 & 0 & 4 & 0 \\ 0 & 0 & 2 & 0 \\ 0 & 0 & 0.5 & 0 \\ 0 & 0 & 0 & 0 \end{pmatrix} = \begin{pmatrix} 1 & 0 & -4 & 0 \\ 0 & 1 & -2 & 0 \\ 0 & 0 & 0.5 & 0 \\ 0 & 0 & 0 & 1 \end{pmatrix}.$$

In the next round, the new basis is the same as the previous one plus ze_2^T, where now

$$z = \begin{pmatrix} 1 \\ 1.5 \\ 0.5 \\ 0 \end{pmatrix} - \begin{pmatrix} 0 \\ 1 \\ 0 \\ 0 \end{pmatrix} = \begin{pmatrix} 1 \\ 0.5 \\ 0.5 \\ 0 \end{pmatrix},$$

and where e_2 is the second unit vector. Thus, its inverse equals

$$
\begin{pmatrix} 1 & 0 & -4 & 0 \\ 0 & 1 & -2 & 0 \\ 0 & 0 & 0.5 & 0 \\ 0 & 0 & 0 & 1 \end{pmatrix} - \frac{1}{1+0.5} \begin{pmatrix} 1 & 0 & -4 & 0 \\ 0 & 1 & -2 & 0 \\ 0 & 0 & 0.5 & 0 \\ 0 & 0 & 0 & 1 \end{pmatrix}
$$

$$
\times \begin{pmatrix} 1 \\ 0.5 \\ 0.5 \\ 0 \end{pmatrix} \begin{pmatrix} 0 & 1 & 0 & 0 \end{pmatrix} \begin{pmatrix} 1 & 0 & -4 & 0 \\ 0 & 1 & -2 & 0 \\ 0 & 0 & 0.5 & 0 \\ 0 & 0 & 0 & 1 \end{pmatrix}
$$

$$
= \begin{pmatrix} 1 & 0 & -4 & 0 \\ 0 & 1 & -2 & 0 \\ 0 & 0 & 0.5 & 0 \\ 0 & 0 & 0 & 1 \end{pmatrix} - \frac{1}{1+0.5} \begin{pmatrix} -1 \\ -0.5 \\ 0.25 \\ 0 \end{pmatrix} \begin{pmatrix} 0 & 1 & -2 & 0 \end{pmatrix}
$$

$$
= \begin{pmatrix} 1 & 2 & -8 & 0 \\ 0 & 2 & -4 & 0 \\ 0 & -0.5 & 1.5 & 0 \\ 0 & 0 & 0 & 1 \end{pmatrix}.
$$

Remark. As a by-product of the above analysis, we have also proved that the new matrix has an inverse, and hence, it is a basic matrix.

Exercise 7.4

Apply the Simplex method to the following LP:

$$
\min_{x_i, 1 \le i \le 6} -x_1 - 2x_2 + x_3 - x_4 - 4x_5 + 2x_6,
$$

$$
\text{s.t.} \quad x_1 + x_2 + x_3 + x_4 + x_5 + x_6 \le 6,
$$

$$
2x_1 - x_2 - 3x_3 + x_4 \le 4,
$$

$$
x_3 + x_4 + 2x_5 + x_6 \le 4,
$$

$$
x_i \ge 0, \quad 1 \le i \le 6
$$

7.6 Degeneracy and Cycling

We have invoked above the assumption of non-degeneracy. Let us ask ourselves what might happen otherwise. The first thing to observe is that in the case where a variable i^* that was selected to be removed from the basis currently has a value of zero, and where $(B^{-1}N)_{i^*j} > 0$, the entrant variable $(x_N)_j$ cannot be increased from zero. The basis can be changed regardless, but — and this is the key point here — both the old and the new bases have the same basic solution. It is only the roles of the variables j and i^*, the former as a non-basic variable and the latter as a basic variable, that are swapped. In particular, the value of the objective function stays unchanged. The main danger is when we move from one basis to another, then to a third one, and so on, without improving the objective function, a cycle might be formed. By this, we mean that a basis will be repeated, and from then on, the algorithm will enter an indefinite loop. In other words, no convergence is guaranteed in the case of degeneracy. Indeed, examples in which cycling occurs have been constructed. Note, however, that this is an unlikely event, even in the presence of degeneracy.

There is another threat in the case of degeneracy: It is possible that the optimality criterion will not be satisfied in an optimal basis. In other words, it is possible that, in spite of the fact that a given basic solution is optimal, a non-basic variable whose corresponding reduced cost is strictly negative will be detected.

Can cycling be avoided? The answer is "yes". Note that in the definition of the Simplex method there is no clear-cut prescription on what to do, first, when more than one non-basic variable can be added to the basis, and second, when more than one basic variable can be removed due to a tie in the ratio test.[k] When looking for cycling prevention rules, we ask ourselves whether there are ways to select the added and/or removed variables in a way that guarantees no cycles. A few such rules have been designed. The simplest one (due to [7]) prescribes always selecting the variable with the lowest index (in both cases of adding and removing variables). Note that

---

[k]In the latter case, the degeneracy of the new basis is guaranteed, as the non-leaving variable in the tie receives a value of zero under the new basic solution.

some order of the variables should be decided and maintained throughout the procedure. See [20, pp. 50–55], for further details.

7.7 The Representation Theorem

Combining Theorem 7.4 with Theorem 7.5 leads to the conclusion that an LP is feasible if and only if it possesses at least one extreme point. On the other hand, as the number of feasible bases is finite, the same is the case with the number of extreme points. Consider the (non-empty) set of extreme points. Clearly, any convex combination of them is also a feasible solution. Now comes the question of whether other feasible solutions exist. We claim, without proof, that no other solutions exist if and only if the feasible set is bounded.

To determine the actual case of an unbounded feasible set, we need to define the concept of a *ray*. Specifically, a non-zero vector $d \in R^n$ is said to be a ray if $d \geq 0$ and $Ad = 0$. It is easy to see that if x is feasible and d is a ray then $x + \alpha d$ is feasible for any scalar $\alpha \geq 0$. See Exercise 7.7. It is also easy to see that the set of rays is a convex set. It is not a bounded set (note, for example, that if d is a ray, then αd is also a ray for any $\alpha \geq 0$), but since it is only the direction of d that matters (and not its magnitude), we can make it bounded by imposing the constraint that $\Sigma_{i=1}^n d_i = 1$. This newly bounded set contains a finite number of extreme points, which are called *extreme rays*.

This leads us to the *representation theorem*, which we state without proof. A detailed proof can be found in [2].

Theorem

Theorem 7.11. *Let x^i, $1 \leq i \leq l^e$, be the set of extreme points of the feasible set of an LP and let d^i, $1 \leq i \leq l^r$, be the corresponding set of extreme rays. Then, $l^e \geq 1$ and $l^r \geq 0$. Also, for any LP feasible solution x, there exist non-negative vectors $\alpha \in R^{l^e}$ and $\mu \in R^{l^r}$ with $\Sigma_{i=1}^{l^e} \alpha_i = 1$ such that*

$$x = \sum_{i=1}^{l^e} \alpha_i x^i + \sum_{i=1}^{l^r} \mu_i d^i.$$

Exercise 7.5

(Caratheodory theorem) Let $x^1, x^2, \ldots, x^k$ be a set of k points in R^n. Let C be the convex hull of these points, namely,

$$C = \left\{ x \in R^n, \ x = \sum_{i=1}^{k} \alpha_i x^i \ \middle| \ \sum_{i=1}^{k} \alpha_i = 1, \ \alpha_i \geq 0, 1 \leq i \leq k \right\},$$

where C is the set of all convex combinations of $x^1, x^2, \ldots, x^k$. Show that any $x \in C$ can be expressed as a convex combination of at most $n + 1$ points[1] out of $x^1, x^2, \ldots, x^k$.

Exercise 7.6

Let X be the set defined by $X = \{x | Ax = b, x \geq 0\}$. The non-zero vector d is said to be a *ray* of $x \in X$ if for any non-negative scalar α, it is the case that $x + \alpha d \in X$.

- Show that d is a ray of $x \in X$ if and only if it is a ray of $y \in X$.
- Show that d is a ray of $x \in X$ if and only if d satisfies $Ad = 0$, and all its entries are non-negative.

Exercise 7.7

Draw the following set X.

$$-3x_1 + x_2 \leq -2,$$

$$-x_1 + x_2 \leq 2,$$

$$-x_1 + 2x_2 \leq 8,$$

$$x_2 \geq 2.$$

- What are the extreme points and extreme rays of X?
- Express $(4, 3)$ as a convex combination of the extreme points plus a non-negative combination of the extreme rays.

---

[1] Of course, these $n + 1$ points vary with the selected x.

Exercise 7.8

The following LP is given:

$$\max_{x_1, x_2, x_3} \ 2x_1 + x_2 - x_3,$$

$$\text{s.t.} \quad x_1 x_2 2x_3 \leq 6,$$

$$x_1 4x_2 - x_3 \leq 4,$$

$$x_i \geq 0, \quad 1 \leq i \leq 3.$$

- What is the set of extreme points of the feasible set?
- What is the (possibly empty) set of extreme rays?
- Express any feasible point as a convex combination of the extreme points coupled a non-negative combination of the extreme rays. Can you now guess the optimal solution?

Chapter 8

Sensitivity Analysis

8.1 Introduction

Sensitivity analysis deals with the question of what happens to the optimal solution and/or to the optimal objective value when one or more of the entries in the data defining the LP change. Changes can be in the cost coefficients, the right-hand side entries, or the constraint matrix itself. Usually the changes are not so large, and hence we try to infer the new problem once the previous one has been solved. Moreover, by a continuity argument, we believe that the changes in the results will not be huge.

The following theorem states the shape of the objective function value as a function of the right-hand side vector or the cost coefficient.

Theorem

Theorem 8.1. *The following two statements hold for a standard LP:*

- *Denote by $f(b)$ the optimal objective value as a function of the right-hand side vector b. Then, $f(b)$ is a convex function.*
- *Denote by $g(c)$ the optimal objective value as a function of the cost coefficient vector c. Then, $g(c)$ is a concave function.*

Proof.

- Denote by $x(b)$ the optimal solution when the right-hand side is b (assuming it exists). Also, let α be with $0 \leq \alpha \leq 1$. It is easy to see that $\alpha x(b^1) + (1 - \alpha)x(b^2)$ is a feasible solution for the right-hand side

$\alpha b^1 + (1 - \alpha)b^2$. Yet, maybe a better solution exists for the latter right-hand side than this one. Hence,

$$\alpha f(b^1) + (1 - \alpha)f(b^2) = c^T(\alpha x(b^1) + (1 - \alpha)x(b^2)) \geq f(\alpha b^1 + (1 - \alpha)b^2),$$

proving that $f(b)$ is convex. Note that in the case where one of the problems with b^1 or b^2 is not feasible, the left-hand side of the above inequality equals infinity, making the correctness of the inequality trivial. Also, note that if one of the the LPs is unbounded, then the same is the case with the mixed right-hand side.

- Denote by $x(c)$ the optimal solution in the case of a cost coefficient vector of c and assume that $0 \leq \alpha \leq 1$. Then,

$$\begin{aligned}
g(\alpha c^1 + (1 - \alpha)c^2) &= (\alpha c^1 + (1 - \alpha)c^2)^T x(\alpha c^1 + (1 - \alpha)x^2) \\
&= \alpha c_1^T x(\alpha c^1 + (1 - \alpha)c^2) + (1 - \alpha)x(\alpha c^1 + (1 - \alpha)c^2) \\
&\geq \alpha c_1^T x(c^1) + (1 - \alpha)x(c^2),
\end{aligned}$$

where the inequality is due to the optimality of $x(c^1)$ and $x(c^2)$. Of course, if one of the three considered problems is infeasible, then the same is the case with the other two. Finally, the above inequality holds trivially in the case where one of the original problems is unbounded. □

You may recall that for a given LP, a basis is optimal if and only if it obeys two conditions. The first is the feasibility condition, namely $B^{-1}b \geq \underline{0}$, and the second is the optimality condition, which requires that the reduced price vector is non-negative, namely $c_N^T - c_B^T B^{-1} N \geq \underline{0}$. Note that the right-hand side vector is needed only to check the first condition, while the cost coefficient vector is needed only for the second condition. Put differently, if B is optimal for some b^1 as a right-hand side vector and satisfies the feasibility condition for another right-hand side vector b^2, then it is also optimal for b^2. The new optimal solution is $x_B = B^{-1}b^2$ with $x_N = \underline{0}$. Likewise, if B is optimal for c^1 and satisfies the optimality condition for c^2, then it is optimal for c^2 too. In fact, in this case, it is also the optimal solution itself (and not only the optimal basis) that does not vary.

Exercise 8.1

A real function f is called *sub-additive* if for any pair of x_1 and x_2, $f(x_1 + x_2) \leq f(x_1) + f(x_2)$. It is called *super-additive* in the case where the reverse inequality holds. Consider the LP

$$\max_x \quad c^T x,$$

$$\text{s.t.} \quad Ax = b,$$

$$x \geq \underline{0}.$$

- Show that the value of the objective function is a sub-additive function of the cost coefficient vector c.
- Show that the value of the objective function is a super-additive function of the right-hand side vector b.

8.2 Changes in the Right-Hand Side

So far, the right-hand side vector has been considered as fixed (like the rest of the data defining the LP). Suppose now we change the right-hand side vector in one way or another, while keeping the rest of the data fixed. Of course, the optimal value of the LP can change accordingly. To show this, denote by $f(b)$ the optimal value when b is the right-hand sidevector. Note that b is the vector of variables in the function f. In other words, $f : R^m \to R$.

> ### Theorem
>
> **Theorem 8.2.** *The set of right-hand side vectors for which B is a feasible basis is convex. Moreover, the set for which it is optimal is also convex.*

Proof. It is easy to see that if $B^{-1}b^1 \geq \underline{0}$ and $B^{-1}b^2 \geq \underline{0}$, then $B^{-1}(\alpha b^1 + (1-\alpha)b^2) \geq \underline{0}$ for any $\alpha \in [0,1]$. Finally, if the optimality condition holds for one b, it also holds trivially for any other b, as the right-hand side vector is not considered in this condition. $\square$

The conclusion from the above theorem is that once an LP is solved, and you then look for some other right-hand side vectors under which the optimal basis does not vary, there is no need to look far: They are nearby. In fact, if $B^{-1}b \gg \underline{b}$, you can move away from b, at least for a while, in any direction and stay feasible. However, once you cross the barrier and continue in the same direction, you will never again get a right-hand side for which B is optimal, as this would contradict convexity. The next result is now immediate.

> **Theorem**
>
> **Theorem 8.3.** *For any b for which a given basis B is optimal, the optimal objective value equals $f(b) = c_B^T B^{-1} b$. In particular, it is a linear function of b at the convex region where B is optimal and the coefficient of b_i is $(c_B^T B^{-1})_i$. Similarly in this region, x_B is a linear function of b: $x_B = B^{-1} b$. In particular, if $(x_B)_i > 0$, then B_{ij}^{-1} is the derivative of $(x_B)_i$ with respect to b_j. Finally, assuming $x_B \gg \underline{0}$, the derivative of $(x_N)_k$ with respect to b_j is zero.*

It should now be clear why $c_B^T B_{.i}^{-1}$ is called the *shadow price* of the ith entry in the right-hand side, $1 \le i \le m$. If the right-hand side presents some commodities that are consumed in order to make a profit represented by the optimal objective value, then the shadow price $c_B^T B_{.i}^{-1}$ is the change in the objective function per unit of a change in the ith entry in b, $1 \le i \le m$. Indeed,

$$\frac{d\,f(b)}{d\,b_i} = c_B^T B_{.i}^{-1}, \tag{8.1}$$

when B^{-1} is the optimal basis at b. Note that this price, or derivative, can be positive as well as negative. We denote it by π_i and it is referred to as the shadow price of the ith entries in the RHS. See Remark 7 on duality at the end of Section 7.3.4.

An interesting case is when the shadow price is zero. Suppose that i is a constraint that was originally put as an inequality and was made to be an equality by the introduction of a slack or a surplus variable. Suppose that this variable is positive in the optimal basis. In particular, it is a basic variable. It is clear that a small change in this entry in the right-hand side will not change the optimal objective value (only the new variable will change slightly, but as its coefficient in the objective function is zero, no effect on the objective function value takes place). Hence, the shadow price should be zero here. Try to see if you can formally prove this fact by proving that $c_B^T B_{.i}^{-1} = 0$ in this case. Hint: $(c_B^T)_i = 0$ and $B^{-1}.i\underline{1} = e_i$, namely it is a unit vector with 1 in position i and zeros elsewhere.

As said above, one can move slightly away from b in any direction and maintain the optimal basis. The question we ask now is how much we can move in the direction e_i and maintain the optimality of the basis that was optimal for b. In fact, since the optimality condition is satisfied, our interest is only in looking for feasibility. Thus, we look for

$$\theta_i^+ = \max\{\theta \geq 0 | B^{-1}(b + \theta e_i) \geq \underline{0}\}.$$

The answer is simple:

$$\theta_i^+ = \min_{j \in B, B_{ji}^{-1} < 0} - \frac{(B^{-1}b)_j}{B_{ji}^{-1}}. \tag{8.2}$$

This is what LP packages sometimes refer to as *the allowable increase*. See Figure 8.3 for an example. Note that if the set where minimization is carried out is empty, i.e., if $B_{ji}^{-1} \geq 0$ for all basic variables j, then one can increase b_i as much as one likes and still keep the optimality of the basis under consideration.

What about θ_i^-, the allowable decrease? The answer is now immediate:

$$\theta_i^- = \min_{j \in B, B_{ji}^{-1} > 0} \frac{(B^{-1}b)_j}{B_{ji}^{-1}}.$$

We have so far dealt with the effect of a change in the right-hand side on the objective function. See (8.1). Such a change first affects the entries in x_B, which in turn changes $c_B^T x_B$, the value of the objective function. So, what is the impact on x_B and where can it be seen? The answer is B^{-1}. Specifically, since $x_B = B^{-1}b$, it follows that x_B is a linear function of b (as long as the basis does not change). As we already observed, the derivative of $(x_B)_i$ with respect to b_j equals B_{ij}^{-1}. In short, B^{-1} tells us all we need to know about the sensitivity of the solution with respect to the right-hand side vector.

What about simultaneous changes in right-hand side entries? It is highly unlikely that if we increase, say, b_i by the allowable increase θ_i^+ and simultaneously decrease, say, b_j by the allowable decrease θ_j^-, we will still get a new right-hand side vector under which B maintains its optimality. What we have here instead is a sufficient condition that is known as the *100–percent rule*. According to this rule, each increase or decrease in any of the right-hand side entries consumes a positive fraction of the total allowable change. Specifically, if the sum of these fractions is less than or equal to one, then B is optimal under the new right-hand side vector. The reasoning behind this is the convexity of the set of right-hand side vectors under which a given basis is optimal. See Theorem 8.2. For an illustration for the case where $m = 2$, see Figure 8.1.

Finally, what happens if b_i increases (for a little bit) beyond the allowable increase stated in (8.2)? One possibility is that the program becomes

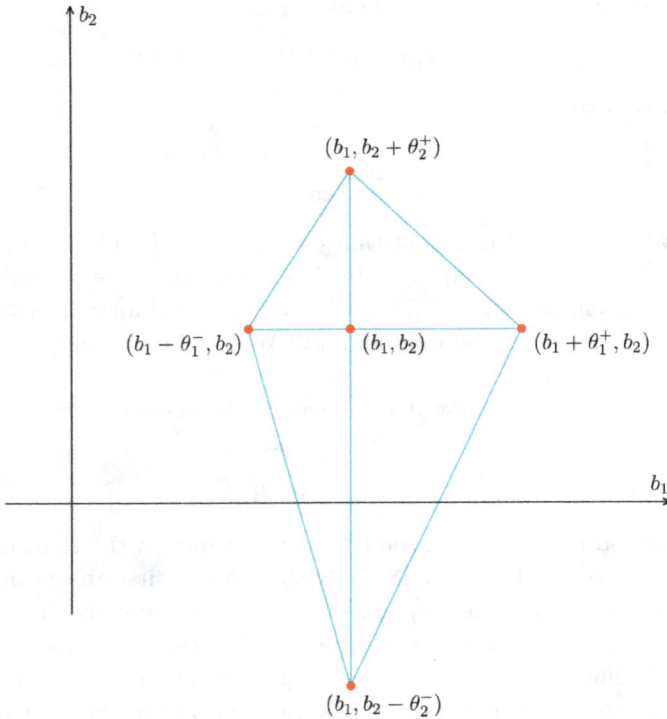

Figure 8.1. 100-percent rule.

infeasible. The other possibility is that another basis becomes optimal. This is a basis adjacent to B, where the removed basic variable is the one where the minimization defining (8.2) is attained (assuming it is unique). In fact, at $b + \theta_i^* e_i$, degeneracy occurs. In particular, two bases are optimal there and both come with one (but different) basic variable whose value is zero. Of course, the optimal objective value is the same. Also, $f(b + \theta e_i)$ as a function of θ is not differentiable at the point $\theta = \theta_i^*$. See Figure 8.2 for a few options. Note that due to the convexity of the function $f(b)$ (see Theorem 8.1), the slope of $f(b + \theta e_i)$ is never decreasing.

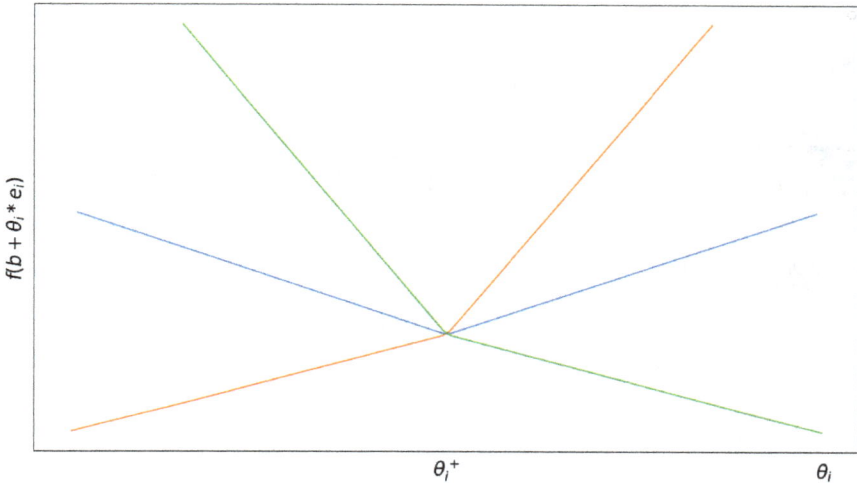

Figure 8.2. Three possibilities for the optimal value as a function of a right-hand side entry.

Example. Consider the following LP:

$$\min_{x_i, 1 \le i \le 4} x_1 + x_2,$$

$$\text{s.t.} \quad x_1 + 3x_2 + x_3 = 30,$$

$$2x_1 + x_3 + x_4 = 20,$$

$$x_i \ge 0, \quad 1 \le i \le 4.$$

It is possible to check that the set of optimal basic variables are $x_B = (x_1, x_2)$. In particular,

$$B = \begin{pmatrix} 1 & 3 \\ 2 & 1 \end{pmatrix} \quad \text{and} \quad B^{-1} = \begin{pmatrix} -0.2 & 0.6 \\ 0.4 & -0.2 \end{pmatrix}.$$

Also,

$$x_B = \begin{pmatrix} x_1 \\ x_2 \end{pmatrix} = \begin{pmatrix} -0.2 & 0.6 \\ 0.4 & -0.2 \end{pmatrix} \begin{pmatrix} 30 \\ 20 \end{pmatrix} = \begin{pmatrix} 6 \\ 8 \end{pmatrix}.$$

	A	B	C	D	E	F	G
1		X1	X2				
2		0	0				
3							
4	b1	1	3	0	<=	30	0
5	b2	2	1	0	<=	20	0
6							
7	Objective value	1	1	0			

Microsoft Excel 16.0 Answer Report
Worksheet: [Book1]Sheet1
Report Created: 22/04/2025 22:15:33
Result: Solver found a solution. All Constraints and optimality conditions are satisfied.
Solver Engine
Solver Options

Objective Cell (Max)

Cell	Name	Original Value	Final Value
$G$7	Objective value	14	14

Variable Cells

Cell	Name	Original Value	Final Value	Integer
$B$2	X1	6	6	Contin
$C$2	X2	8	8	Contin

Constraints

Cell	Name	Cell Value	Formula	Status	Slack
$G$4	<=	30	$G$4<=$F$4	Binding	0
$G$5	<=	20	$G$5<=$F$5	Binding	0
$B$2	X1	6	$B$2>=0	Not Binding	6
$C$2	X2	8	$C$2>=0	Not Binding	8

Microsoft Excel 16.0 Sensitivity Report
Worksheet: [Book1]Sheet1
Report Created: 22/04/2025 22:15:33

Variable Cells

Cell	Name	Final Value	Reduced Cost	Objective Coefficient	Allowable Increase	Allowable Decrease
$B$2	X1	6	0	1	1	0.666666667
$C$2	X2	8	0	1	2	0.5

Constraints

Cell	Name	Final Value	Shadow Price	Constraint R.H. Side	Allowable Increase	Allowable Decrease
$G$4	<=	30	0.2	30	30	20
$G$5	<=	20	0.4	20	40	10

	A	B	C	D	E	F	G
1		X1	X2				
2		0	0				
3							
4	b1	1	3	0	<=	60.1	0
5	b2	2	1	0	<=	20	0
6							
7	Objective value	1	1	0			

Microsoft Excel 16.0 Answer Report
Worksheet: [Book1]Sheet1
Report Created: 22/04/2025 22:22:35
Result: Solver found a solution. All Constraints and optimality conditions are satisfied.
Solver Engine
Solver Options

Objective Cell (Max)

Cell	Name	Original Value	Final Value
$G$7	Objective value	0	20

Variable Cells

Cell	Name	Original Value	Final Value	Integer
$B$2	X1	0	0	Contin
$C$2	X2	0	20	Contin

Constraints

Cell	Name	Cell Value	Formula	Status	Slack
$G$4	<=	60	$G$4<=$F$4	Not Binding	0.1
$G$5	<=	20	$G$5<=$F$5	Binding	0
$B$2	X1	0	$B$2>=0	Binding	0
$C$2	X2	20	$C$2>=0	Not Binding	20

Microsoft Excel 16.0 Sensitivity Report
Worksheet: [Book1]Sheet1
Report Created: 22/04/2025 22:22:35

Variable Cells

Cell	Name	Final Value	Reduced Cost	Objective Coefficient	Allowable Increase	Allowable Decrease
$B$2	X1	0	-1	1	1	1E+30
$C$2	X2	20	0	1	1E+30	0.5

Constraints

Cell	Name	Final Value	Shadow Price	Constraint R.H. Side	Allowable Increase	Allowable Decrease
$G$4	<=	60	0	60.1	1E+30	0.1
$G$5	<=	20	1	20	0.033333333	20

Figure 8.3. Excel's solver output for LP1 (left) and LP2 (right).

What is the allowable increase for b_1? The answer (see (8.2)) is

$$\theta_1^+ = \min_{j \in B, B_{j1}^{-1} < 0} -\frac{(B^{-1}b)_j}{B_{j1}^{-1}} = \min\left\{-\frac{6}{-0.2}\right\} = 30.$$

Change now b_1 to 59.9, which is within the range of the allowable increase. Now,

$$x_B = \begin{pmatrix} x_2 \\ x_3 \end{pmatrix} = \begin{pmatrix} -0.2 & 0.6 \\ 0.4 & -0.2 \end{pmatrix} \begin{pmatrix} 59.9 \\ 20 \end{pmatrix} = \begin{pmatrix} 0.02 \\ 19.96 \end{pmatrix}.$$

Observe that the basic variables x_1 is now close to zero. Finally, increase b_1 to 60.1, which is outside the allowable increase. It is possible to check that now $x_B = (x_2, x_3)$. Then,

$$B = \begin{pmatrix} 3 & 1 \\ 1 & 0 \end{pmatrix} \quad \text{and} \quad B^{-1} = \begin{pmatrix} 0 & 1 \\ 1 & -3 \end{pmatrix}.$$

Also,

$$x_B = \begin{pmatrix} 0 & 1 \\ 1 & -3 \end{pmatrix} \begin{pmatrix} 60.1 \\ 20 \end{pmatrix} = \begin{pmatrix} 20 \\ 0.1 \end{pmatrix}.$$

Note that now the previously non-basic variable (now basic) gets a value close to zero.

8.2.1 *A note on the case of degeneracy*

All of the above breaks down in the presence of degeneracy. In particular, in this case, it is possible that a few bases, say two, are optimal in the same right-hand side vector b. Note, however, that the optimal solution is the same under both bases: The only change is the classification of some (usually two) variables, whose values in this solution equal zero, swapping the roles of basic and non-basic variables. Yet, despite of the fact that the optimal solutions is the same under the two bases, the derivatives stated in (8.1) are different here. Still, it is possible that a small perturbation in the right-hand side vector will break the tie between these bases. Consider, for example, the case where increasing b_i by a bit makes one basis uniquely optimal, while decreasing it a bit makes the other one uniquely optimal. Loosely speaking, the question becomes: Which of the two derivatives should be taken? The answer is none of the above, or better put, at b, $f(b)$ is not differentiable. This should not be confused with the fact that $f(b)$ is still continuous in b.

8.3 Changes in the Cost Coefficients

Suppose that an LP is solved for some given data. In particular, an optimal basis is found. Suppose now that the cost coefficients are undergoing a change. The question we pose here is whether the original optimal basis maintains its optimality. Note that we need to consider only the optimality condition, as the feasibility condition is guaranteed to be met as no changes were performed at the right-hand side. As importantly, note that since the change is in the cost coefficients, then if the basis stays optimal, the optimal solution $x_B = B^{-1}b$ does not change too. Only the objective function value, $c_B^T B^{-1}b$, will change in the case where the change in c effects c_B^T.

The first thing to observe is that, here too, the set of cost vectors under which a given basis is optimal is convex. Hence, we proceed to the question of allowable increase or decrease in each entry individually (when all other entries are kept unchanged).

The analysis is simple if the change takes place in a non-basic variable: There is no limit on how much its coefficient can increase, while its (non-negative) reduced cost coincides with the amount by which this coefficient decreases. Indeed, decreasing the cost coefficient further makes the reduced cost negative, and hence as a result, this variable is now added to the basis, and the basis changes.

If the change is in a basic variable, the analysis should follow the same lines of our above analysis regarding a change in the right-hand side vector. Specifically, a change from c_B^T to $c_B^T + \beta e_i$, with $\beta > 0$, does not affect the optimality of the basis if and only if for **all** non-basic variables, the reduced cost stays non-negative, namely

$$c_j - (c_B^T + \beta e_i)B^{-1}N_{.j} \geq 0, \quad j \in N$$

or, equivalently,

$$\bar{c}_j - \beta(B^{-1}N)_{ij} \geq 0, \quad j \in N,$$

where $\bar{c}_j$ is the reduced cost of the non-basic variable $j \in N$. In other words, the allowable increase is

$$\beta_i^+ = \min_{j,(B^{-1}N)_{ij}>0} \frac{\bar{c}_j}{(B^{-1}N)_{ij}}.$$

Note that the allowable increase can be unbounded in the case where the above minimization is over an empty set. It is notable that if the allowable increase is infinite, namely no matter how costly it becomes, this variable still needs to be in the optimal basis. In particular, it needs to be positive. Of course, we are not satisfied with this outcome and would like to remove this variable from the optimal basis once its cost is huge. Yet, this seems impossible. The conclusion is that the current value of this variable (recall that it does not vary with c) is the lowest possible value among all feasible solutions.[a]

---

[a]It is possible to put this in algebraic terms. Specifically, recall the relationship $(x_B)_i = (B^{-1}b)_i - (B^{-1}N)_{ij}x_j$, which was used when we considered adding the variable x_j into the basis. Assuming that $(B^{-1}N)_{ij} < 0$ and reversing the relationship, we can see that if $(x_B)_i$ decreases (a plus when its cost coefficient increases), then sooner or later, x_j becomes negative. However, this is not the only option (since it comes with keeping all other non-basic variables at zero). Yet, since this happens to **all** non-basic variables when $\beta_i^+ = \infty$, any other direction also leads to a reduction in the values of some or all non-basic variables. This is the case since any other direction is a positive

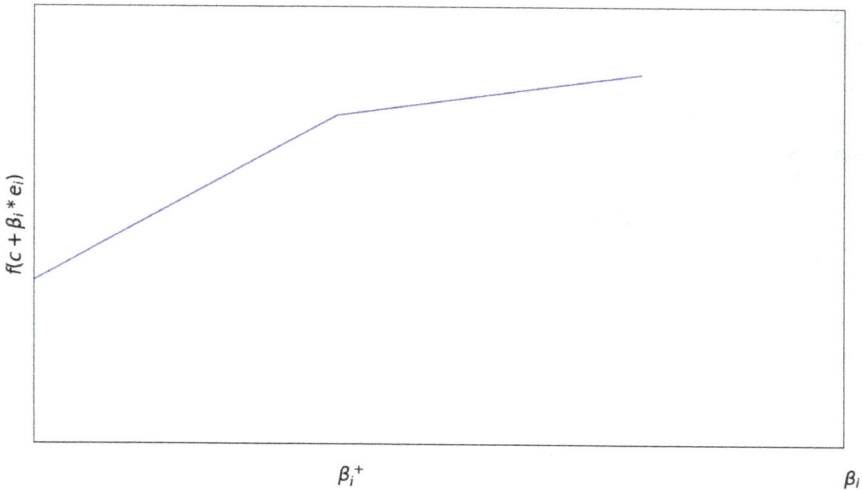

Figure 8.4. The optimal value as a function of c_i.

A similar expression exists for the allowable decrease β_i^-:

$$\beta_i^- = \min_{j,(B^{-1}N)_{ij}<0} \frac{\bar{c}_j}{-(B^{-1}N)_{ij}}.$$

From the above, we can infer that if one looks at the optimal objective function value as a function of the cost parameters, then its derivative with respect to a parameter that corresponds to a non-basic variable is zero. Also, the derivative with respect to a basic variable x_i is $(B^{-1}b)_i = \sum_{j \in B} B_{ij}^{-1} b_j$.

You may recall the 100-percent rule which was applied on changes in the right-hand side vector. It is applied verbatim to changes in the cost coefficients. In fact, it is applied to changes in the vector (b, c) where some of the changes are at the right-hand side, while others are at the cost coefficients.

What happens in the case where β_i is increased beyond β_i^+ for a basic variable $(x_B)_i$? There is no question regarding the feasibility of the LP. In particular, the optimal basic solution stays feasible. Had it stayed optimal, the objective function would have been linear with β_i. As this is not

---

linear combination of the $n - m$ elementary directions. This, of course, will violate the feasibility of the solution, and hence such directions are not feasible.

the case, the slope of the optimal objective function now becomes smaller. Moreover, as argued for θ_i^*, the objective value is not differentiable at β_i^+. See Figure 8.1 for some cases. An alternative argument can utilize the fact that the optimal objective value is concave with respect to its cost coefficients. See Theorem 8.1.

Suppose that you have solved a feasible LP (regardless of whether you end up with an optimal solution or with an indication that the LP is unbounded). Then suppose that you are asked to solve a similar LP but with a different cost vector. What should you do? First, you can check the signs of the reduced costs in the new vector. If all of them are non-negative, you are done: Whatever your conclusion on the previous LP was, it stays for the new LP. Specifically, if it was an unbounded LP before, it is still unbounded, or if it has an optimal basis, the same optimal basis stays optimal (maybe with a new value for the objective function). Otherwise, namely if one of the new reduced costs becomes negative, you at least have in hand a feasible basis (the terminating one from the simplex algorithm), and you can restart the application of the simplex algorithm to the new LP.

Our final application of this sensitivity analysis deals with the introduction of a new variable. Denote by c_j the variable's cost coefficient and by A^j its constraint column. Suppose that the original LP without this variable solved. This implies that currently we hold the best solution, out of those in which the new variable receives a value of zero. Suppose that we decide to increase this variable from zero by a bit, say by ϵ. The value of objective function changes not only by $c_j\epsilon$ (which is immediate) but also by $-\epsilon\Sigma_{i=1}^{m}\pi_i A_{ij}$, where $\pi_i = c_B^T B^{-1}.i$ is the shadow price that corresponds to constraint i, $1 \le i \le m$. Indeed, due to the change at the other basic variables, some of the right-hand side is consumed by the new variable. In summary, a gain (in terms of a reduction in the resulting optimal LP value) can be realized if and only if $c_j - \Sigma_{i=1}^{m}\pi_i A_{ij} < 0$. Thus, this is the criterion with which we decide whether or not to introduce this new variable. But reconsider the definition of the shadow prices π and get that the criterion is that the variable should be introduced if and only if

$$c_j - \sum_{i=1}^{m} c_B^T B_{.i}^{-1} A_{ij} < 0.$$

Do you recognize this figure? Yes, it is the reduced cost of the non-basic variable j with respect to the final basis. In fact, we could have answered the posed question by inspecting the reduced cost of the new variable. Finally, note that if we decide to introduce this new variable, the road to optimality is open. The new variable is added to the basis, and the Simplex method continues from this point on. Note that the new variable will never be removed from the basis (as the optimal value when this variable was effectively set to zero was already in hand before the Simplex algorithm restarted).

Exercise 8.2

For the example given in Section 7.3.5, answer the following questions:

1. What are the shadow (dual) prices for any of the right-hand side entries?
2. For the first two right-hand side entries, determine what are the ranges for which these entries can vary (when all other coefficients are fixed at their original values) while the optimal basis is maintained.
3. (Cont.) Give an explicit expression for the optimal solutions and for the optimal objective function values in these ranges.
4. What is the optimal solution and what is the value of the objective function when the first entry in the right-hand side is changed from 48 to 30, and to 60?
5. What is the optimal solution and what is the value of the objective function when the second entry in the right-hand side is changed from 20 to 22, and to 18?
6. Suppose a new variable (say, shelves) is suggested. Its price coefficient is 10 and its constraint column is $(2, 1, 0.5, 0)^T$. Should any production of shelves be initialized?
7. Suppose the right-hand side is changed to $(55, 19, 8, 5)$. Does the optimal basis change? If not, what is the new optimal solution and what is the new objective function value?
8. For each of the non-basic cost coefficients find the range in which they can vary (when all other coefficients are fixed in their original values) while maintaining the optimal solution.
9. (Cont.) Give an expression for the optimal solution and for the optimal objective function value. Hint: The answer is trivial.

10. For the non-slack variable in the basis with the lowest index, find what is the range in which its cost coefficient can change while maintaining the optimality of the optimal basis.

Exercise 8.3

The following LP is given:

$$\min_x \quad c^T x,$$
$$\text{s.t.} \quad Ax \geq b,$$
$$x \geq \underline{0},$$

with x^* as an optimal solution. Let A_1, A_2, b_1, b_2 be defined as follows

$$A = \begin{pmatrix} A_1 \\ A_2 \end{pmatrix} \quad \text{and} \quad b = \begin{pmatrix} b_1 \\ b_2 \end{pmatrix}.$$

It is given that $A_1 x^* = b_1$ and $A_2 x^* > b_2$. Show that x^* is also an optimal solution to the following two LPs:

$$\min_x \quad c^T x,$$
$$\text{s.t.} \quad A_1 x \geq b_1,$$
$$x \geq \underline{0},$$

and

$$\min_x \quad c^T x,$$
$$\text{s.t.} \quad A_1 x = b_1,$$
$$x \geq \underline{0}.$$

Exercise 8.4

Consider a standard LP which possesses an optimal basic solution. Let x_j be a non-basic variable there. Assume that the reduced cost of x_j is positive. Denote by N^j its corresponding column and by N_{kj} an entry in this column. The following question deal with sensitivity analysis with respect to these entries.

1. Argue that this optimal basic solution stays feasible for any change in all entries in N^j.

2. What is the allowable increase in N_{kj} which keeps this basic solution optimal? (use the notation introduced in class).
3. What is the allowable decrease?
4. Argue that at least one of the above two values is infinite. When is this the case for both of them?
5. Exemplify the 100-percent rule when dealing with simultaneous changes in entries of N^j. In particular, explain why it works.

Chapter 9

Solutions to Exercises

9.1 Chapter 1

Exercise 1.1

Let $Y_1, Y_2, \ldots, Y_n$ be a series of n numbers.

1. Prove that the arithmetic mean of the series minimizes the function

$$\sum_{i=1}^{n} (Y_i - y)^2$$

   with respect to y.

2. Assume further that $Y_i > 0$, $1 \leq i \leq n$. Prove that the geometric mean of the series minimizes the function

$$\sum_{i=1}^{n} (\log Y_i - \log y)^2$$

   with respect to y.

3. Prove that the harmonic mean of the series minimizes the function

$$\sum_{i=1}^{n} \left(\frac{1}{Y_i} - \frac{1}{y} \right)^2$$

   with respect to y.

Solution:

1.

$$\sum_{i=1}^{n}(Y_i - y)^2 = \sum_{i=1}^{n}Y_i^2 - 2y\sum_{i=1}^{n}Y_i + ny^2.$$

This is a quadratic function in y, and its minimum (as shown in the text) is attained at the point

$$\frac{2\sum_{i=1}^{n}Y_i}{2n} = \overline{Y},$$

where $\overline{Y}$ is the arithmetic mean of the series.

2. From the previous item, we infer that the $\log y$ that minimizes

$$\sum_{i=1}^{n}(\log Y_i - \log y)^2$$

is

$$\log y = \overline{\log Y}.$$

Hence, the optimizer y is

$$y = e^{\overline{\log Y}} = e^{\frac{1}{n}\sum_{i=1}^{n}\log Y_i} = \left(e^{\sum_{i=1}^{n}\log Y_i}\right)^{1/n}$$

$$\left(e^{\log\prod_{i=1}^{n}Y_i}\right)^{1/n} = \left(\prod_{i=1}^{n}Y_i\right)^{1/n},$$

which is the geometric mean.

3. Using the solution of (1) but for the series $1/Y$, we conclude that the minimum of

$$\sum_{i=1}^{n}\left(\frac{1}{Y_i} - \frac{1}{y}\right)^2$$

is attained when

$$\frac{1}{y} = \overline{\frac{1}{Y}}.$$

In other words,

$$\frac{1}{y} = \frac{1}{n}\sum_{i=1}^{n}\frac{1}{Y_i}.$$

Thus,

$$y = \frac{n}{\sum_{i=1}^{n} \frac{1}{Y_i}},$$

which is the harmonic mean of the series.

Exercise 1.2

Suppose that X follows a Poisson distribution with parameter λ, namely $P(X = i) = e^{-\lambda} \cdot \frac{\lambda^i}{i!}$ for any integer i, $i \geq 0$.

1. Show that $M_X(s) = e^{\lambda(e^s - 1)}$.
2. Use the Chernoff bound to show that for $a > \lambda$, $P(X \geq a) \leq e^{a-\lambda}(\frac{\lambda}{a})^a$.

Solution:

1.

$$M_X(s) = \sum_{k=0}^{\infty} e^{sk} e^{-\lambda} \frac{\lambda^k}{k!} = e^{-\lambda} \sum_{k=0}^{\infty} \frac{(\lambda e^s)^k}{k!} = e^{-\lambda} e^{\lambda e^s} = e^{\lambda(e^s - 1)},$$

as required.

2. Using the moment-generating function just derived, and invoking the Chernoff bound, we infer that

$$P(X > a) \leq \min_{s > 0} e^{\lambda(e^s - 1) - sa}.$$

By taking the derivative of the exponent and setting it to zero, we observe that the above inequality is minimized at $s = \log(a/\lambda)$, which is positive as $a > \lambda$. Inserting this value for s in the above inequality, we conclude that

$$P(X > a) \leq e^{a-\lambda} \left(\frac{\lambda}{a}\right)^a.$$

Exercise 1.3

A non-negative random variable is said to follow an exponential distribution with parameter λ, if its density function at the point $x \geq 0$ equals $\lambda e^{-\lambda x}$.

1. Based on a random sample of such random variables, what is the likelihood function?
2. What is the maximum likelihood estimator for λ?

Solution:

1. For a random sample X_i, $1 \leq i \leq n$, whose members follow an exponential distribution with parameter λ, the likelihood function equals

$$L(X_1, \ldots, X_n; \lambda) = \prod_{i=1}^{n} \lambda e^{-\lambda X_i} = \lambda^n e^{-\lambda \sum_{i=1}^{n} X_i}.$$

2. The logarithm of the likelihood function is

$$\log L(X_1, \ldots, X_n; \lambda) = n \log \lambda - \lambda \sum_{i=1}^{n} X_i.$$

Taking the derivative with respect to λ and setting it to zero implies that the maximum likelihood estimator for λ, denoted below by $\hat{\lambda}$, satisfies

$$\frac{n}{\hat{\lambda}} - \sum_{i=1}^{n} X_i = 0.$$

Hence,

$$\hat{\lambda} = \frac{n}{\sum_{i=1}^{n} X_i} = \frac{1}{\overline{X}}. \tag{1.1}$$

where $\overline{X}$ is the sample mean.

9.2   Chapter 2

Exercise 2.1

Let $Q \in R^{n \times n}$ be a symmetric matrix and let (λ_i, ω_i), $1 \leq i \leq n$, be the set of pairs of eigenvalues and their corresponding eigenvectors.[a] The vector $x \in R^n$ is called positively shaped if $x^T Q x \geq 0$. Define a negatively shaped vector accordingly.

(a) Show that ω_i is positively shaped if and only if $\lambda_i > 0$.
(b) Show that the set of positively shaped eigenvectors spans a linear subspace.
(c) Is the set of positively shaped vectors a convex set? Prove your answer or give a counterexample.

---

[a]The existence of eigensystems in the case of symmetric matrices is established in Example 6 in Section 4.2.1.

Solution:

(a) $\omega_i^T Q \omega_i = \omega_i^T \lambda_i \omega_i = \lambda_i ||\omega_i||^2$. Note that $||\omega_i|| > 0$. The whole term is therefore positive if and only if $\lambda_i > 0$.

(b) Suppose, without loss of generality, that only $k \leq n$ eigenvectors are positively shaped, indexed by $\omega_1, \ldots, \omega_k$. We then need to show that

$$\left(\sum_{i=1}^k \alpha_i \omega_i \right)^T Q \left(\sum_{i=1}^k \alpha_i \omega_i \right) \geq 0$$

for any set of coefficients $\alpha_1, \ldots, \alpha_k$. Indeed,

$$\left(\sum_{i=1}^k \alpha_i w_i \right)^T Q \sum_{i=1}^k \alpha_i w_i = \sum_{i=1}^k \alpha_i^2 w_i^T Q w_i + \sum_{i=1}^n \sum_{1 \leq j \neq i \leq n} \alpha_i \alpha_j w_i^T Q w_j.$$

The first of the two terms above is non-negative by definition, while the second term equals zero since where $1 \leq j \neq i \leq n$, $w_i^T Q w_j = \lambda_i w_i^T w_j = 0$, due to the orthogonality of eigenvectors of a symmetric matrix. See, e.g., [14, p. 139].

(c) No. For example:

$$Q = \begin{pmatrix} 1 & 0 \\ 0 & -1 \end{pmatrix}.$$

Then $v^T Q v = v_1^2 - v_2^2$. The vectors $(1,0), (-1, \frac{1}{2})$ are both positively shaped, but the midpoint $(0, \frac{1}{4})$ is negatively shaped.

Exercise 2.2

Consider the following function $f : R^2 \to R$[b]:

$$f(x,y) = \frac{x+y}{x^2 + y^2 + 1}.$$

(a) What is the gradient of this function?

(b) What are the two stationary points of f?

(c) What are the values of the function at the stationary points? Which value is a candidate for the global maximum and which one is for the minimum?

---

[b]This equation appears in [3, p. 28].

(d) Show that for any (x, y), $f(x, y)$ is between the two values derived in the previous item.

Hint: Argue that $f(x, y) \leq \sqrt{2} \frac{\sqrt{x^2 + y^2}}{x^2 + y^2 + 1}$. Now use the fact that the right-hand side is a single value function (of $\sqrt{x^2 + y^2}$) whose global maximum equals $\sqrt{2}/2$.

(e) Conclude that the two stationary points are indeed extreme points. Why was the previous item necessary?

Solution:

(a) $\nabla f = \frac{1}{(x^2 + y^2 + 1)^2} \begin{bmatrix} -x^2 + y^2 + 1 - 2xy \\ x^2 - y^2 + 1 - 2xy \end{bmatrix}$.

(b) Setting the gradient to zero, i.e., $-x^2 + y^2 + 1 - 2xy = x^2 - y^2 + 1 - 2xy = 0$, leads to $x^2 - y^2 = 0$ and $2xy = 1$. Thus, the absolute values of x and y are identical, and $xy = \frac{1}{2}$, leading to only two possibilities: $x = y = \frac{1}{\sqrt{2}}$ or $x = y = -\frac{1}{\sqrt{2}}$.

(c) $f(x = y = \frac{1}{\sqrt{2}}) = \frac{1}{\sqrt{2}}$, $f(x = y = -\frac{1}{\sqrt{2}}) = -\frac{1}{\sqrt{2}}$. The first point is a candidate for the global maximum, and the second is a candidate for the global minimum.

(d) We need to show that $-\frac{1}{\sqrt{2}} \leq \frac{x+y}{x^2 + y^2 + 1} \leq \frac{1}{\sqrt{2}}, \forall x, y$. The right hand-side inequality is equivalent to $0 \leq x^2 + y^2 + 1 - \sqrt{2}(x + y)$. Rewriting the right-hand side term as $(x - \frac{1}{\sqrt{2}})^2 + (y - \frac{1}{\sqrt{2}})^2$, we can see now that it is non-negative. In a similar manner, we can show that this holds also for the left-hand side inequality.

(e) We conclude that the points from item (c) are indeed the global maximum and minimum, since their objective values are the largest and smallest values of the function (as proved in (d)).

Exercise 2.3

Consider the function[c] $f : R^3 \to R$ defined as $f(x_1, x_2, x_3) = x_1(x_2 - 1) + x_3(x_3^2 - 3)$.

(a) Compute the gradient ∇f.
(b) Show that both $(0, 1, 1)^T$ and $(0, 1, -1)^T$ are stationary points of f. Are there any other stationary points?
(c) Show that none of the points above are the local minima or maxima.

---

[c]This question is posed in [10, p. 114].

Solution:

(a) $\nabla f = (x_2 - 1, x_1, 3x_3^2 - 3)^T$.

(b) It can be easily shown that the above gradient vanishes at the points $(0, 1, 1)$ and $(0, 1, -1)$. Moreover, it is the unique point with this property.

(c) Define the function $g(\epsilon) = f(0, 1, 1+\epsilon)$. It is possible to see that $g'(\epsilon) = 3\epsilon^2 + 6\epsilon$, such that its signs, for small values for ϵ, coincide with that of ϵ. Thus, in the direction $d = (0, 1, 1)$ the function increases, while in the direction $-d = (0, 0, -1)$ it decreases. The conclusion is that $(0, 1, 1)$ is a reflection point. A similar argument holds for the point $(0, 1, -1)$.

Exercise 2.4

Show that if x_{k+1} is derived from x_k based on an exact line search when the steepest descent method is applied, then the gradients of f at x_k and at x_{k+1} are orthogonal.

Solution:

The steepest descent update is $x_{k+1} = x_k - \alpha^* \nabla f(x_k)$, where $\alpha^* = \arg\min_\alpha f(x_k - \alpha \nabla f(x_k))$. In particular,

$$\frac{d}{d\alpha} f(x_k - \alpha \nabla f(x_k))|_{\alpha=\alpha^*} = 0.$$

But by the chain rule,

$$\frac{d}{d\alpha} f(x_k - \alpha \nabla f(x_k)) = -\alpha \nabla f(x_k)^T \nabla f(x_k - \alpha x_k),$$

which as $\alpha = \alpha^*$ equals zero. Hence,

$$\nabla f(x_k)^T \nabla f(x_k - \alpha x_k) = \nabla f(x_k)^T \nabla f(x_{k+1}) = 0.$$

Suppose $\alpha^* > 0$ (otherwise, if $\alpha^* = 0$, then we know that the current x_k is a local minimum, and the gradient $\nabla f(x_k)$ is zero). Then α^* is a local minimum, and therefore

$$\frac{d}{d\alpha} f(x_k - \alpha \nabla f(x_k))|_{\alpha^*} = 0. \tag{2.2}$$

On the other hand,

$$\frac{d}{d\alpha} f(x_k - \alpha \nabla f(x_k)) = \frac{d}{d\alpha} (-\alpha \nabla f(x_k))^T \nabla f(x_k - \alpha \nabla f(x_k)).$$

Using $\frac{d}{d\alpha}(-\alpha\nabla f(x_k)) = -\nabla f(x_k)$ and $x_{k+1} = x_k - \alpha\nabla f(x_k)$, the derivative equals $\nabla f(x_k)^T\nabla f(x_{k+1})$, which, by (2.2), vanishes at α^*. Thus, consecutive steepest descent directions are orthogonal.

Exercise 2.5

Let $A \in R^{m\times n}$, where $m > n$. Assume that it is a full-rank matrix.

(a) What is the rank of A?
(b) Prove that both $A^T A$ and AA^T are symmetric.
(c) What is the rank of each of the above matrices? Which one, if any, is invertible?

Solution:

(a) The rank of A is n.
(b) $(AA^T)^T = (A^T)^T A^T = AA^T$, $(A^T A)^T = A^T(A^T)^T = A^T A$.
(c) Both have rank n. Since $AA^T \in R^{n\times n}$, it is invertible (its rank is equal to the number of rows/columns), whereas $A^T A \in R^{m\times m}$, and therefore it is non-invertible (its rank is less than the number of rows/columns).

Exercise 2.6

Consider the minimization of the quadratic function $f(x) = \frac{1}{2}x^T Qx - b^T x$.

(a) What is $\nabla f(x)$? What is the steepest descent direction in x?
(b) Denote by d the steepest descent direction in x. Show that

$$\frac{d^T d}{d^T Qd} = \arg\min_{\alpha\in R} f(x + \alpha d).$$

(c) Let d_k and d_{k+1} be two consecutive directions generated by the steepest descent method for solving f(x). Show that d_k and d_{k+1} are orthogonal.

Solution:

(a) The gradient is $Qx - b$, and therefore the steepest descent direction is obtained by negating the gradient.
(b) $f(x + \alpha d) = \frac{1}{2}(x + \alpha d)^T Q(x + \alpha d) = \frac{1}{2}\alpha^2 d^T Qd + \alpha d^T Qx + \frac{1}{2}x^T Qx - b^T(x + \alpha d)$. Differentiate with respect to α: $f'(x + \alpha d) = \alpha d^T Qd + d^T Qx - b^T d$, then its root becomes $\alpha = \frac{b^T d - d^T Qx}{d^T Qd}$. The numerator is

$-d^T(Qx - b) = d^T d$, yielding the term

$$\alpha = \frac{d^T d}{d^T Q d}.$$

(c) Denote $d_k = -(Qx_k - b), d_{k+1} = -(Qx_{k+1} - b)$. We can now derive an expression of x_{k+1} using the steepest descent method: $x_{k+1} = x_k + \frac{d_k^T d_k}{d_k^T Q d_k} d_k$ (using previous result (b)). Plugging that into d_{k+1}, we get that $d_{k+1} = -(Q(x_k + \frac{d_k^T d_k}{d_k^T Q d_k} d_k) - b) = -(-d_k + Q \frac{d_k^T d_k}{d_k^T Q d_k} d_k)$. The dot product between two consecutive directions is therefore

$$d_k^T d_{k+1} = d_k^T \left(-d_k + Q \frac{d_k^T d_k}{d_k^T Q d_k} d_k \right) = -d_k^T d_k + d_k^T Q d_k \frac{d_k^T d_k}{d_k^T Q d_k} = 0.$$

Note that this is a special case of Exercise 2.4.

Exercise 2.7

Let $f(x) : R \to R$. We look for the polynomial $p(x)$ of degree n that solves the following optimization problem:

$$\min_{p(x)} \int_0^1 (f(x) - p(x))^2 dx.$$

Show that it is an unconstrained quadratic programming problem with $n+1$ decision variables (see previous exercise for a definition). In particular, find the Q matrix, the b vector, and the free coefficient.

Solution:

Denote $p(x) = \sum_{i=0}^n \alpha_i x^i$. This polynomial can be written as a scalar product $\alpha^T \tilde{x}$ for $\alpha, \tilde{x} \in R^{n+1}$, where α_i is the ith term in α and $\tilde{x}_i = x^i$, $0 \le i \le n$. This leads to

$$\min_\alpha \int_0^1 (f^2(x) - 2f(x)\alpha^T \tilde{x} + \alpha^T \tilde{x}\tilde{x}^T \alpha) dx.$$

This is a quadratic function $x^T Q x + b^T x + c$, where the free coefficient c is $\int_0^1 f^2(x) dx$, the matrix $Q \in R^{(n+1)\times(n+1)}$ such that $Q_{k,\ell} = \int_0^1 x^k x^\ell dx$, $0 \le k, \ell \le n$, and the vector $b \in R^{n+1}$ such that $b_k = -2\int_0^1 f(x)x^k dx$, $0 \le k \le n$.

Exercise 2.8

Consider the following unconstrained quadratic program:

$$\min_{x \in R^n} \frac{1}{2} x^T Q x - b^T x$$

for some square, symmetric, and positive matrix Q. Let $v_1, \ldots, v_n$ be a set of orthogonal eigenvectors of Q. Suppose that for some given x_0, the gradient at that point belongs to the subspace spanned by $v_1, \ldots, v_m$: $Sp\{v_1, \ldots, v_m\}$ for some $m < n$. Suppose that an iteration of the steepest descent algorithm was executed on x_0. Denote by x' the resulting vector.

(a) Show that the gradient of the objective function at the new point x' also belongs to the subspace $Sp\{v_1, \ldots, v_m\}$.
(b) State whether this is a strong or a weak property of the steepest descent algorithm.

Solution:

(a) The gradient at the point x_0 is $Qx_0 + b$. Thus, the point reached at the end of the current iteration is

$$x' = x_0 - \alpha^*(Qx_0 + b). \tag{2.3}$$

In the next iteration, we take the gradient $Qx' + b$ at x', which equals

$$Qx_0 + b - \alpha^* Q(Qx_0 + b) \tag{2.4}$$

(obtained by inserting the term in (2.3)). This term includes $Qx_0 + b$, which is assumed to lie in the subspace spanned by $v_1, \ldots, v_m$. Since $v_1, \ldots, v_m$ are the eigenvectors of Q, it follows that $Q\left(\sum_{i=1}^m \beta_i v_i\right) = \sum_{i=1}^m \beta_i' v_i$, and in particular $Q(Qx_0 + b) \in Sp\{v_1, \ldots, v_m\}$ as we have assumed that $Qx_0 + b \in Sp\{v_1, \ldots, v_m\}$. Thus, the gradient at x' (see (2.4)) is the sum of two vectors in $Sp\{v_1, \ldots, v_m\}$. Hence, this point lies in the same subspace as well.

(b) This is actually a weak property, since in this case, the steepest descent path is limited to a mere subspace. Denote by x^* the optimal solution. As there is no guarantee that $x^* - x_0$ belongs to this subspace (and usually it does not) then the path will not lead to the desired point.

Exercise 2.9

Let $g_i(x)$, $1 \le i \le n$, be n functions from R^n to itself. Our goal is to find a solution $x \in R^n$ that satisfies $g_i(x) = 0$, $1 \le i \le n$.

(a) Show that this problem is equivalent to solving the unconstrained optimization problem

$$\min_{x \in R^n} \sum_{i=1}^n g_i^2(x).$$

(b) Describe how the steepest descent method is applied to this optimization problem.
(c) Describe how Newton's method is applied to this optimization problem.
(d) How do you rate replacing all of the above where the objective function to be minimized is

$$\min_{x \in R^n} \sum_{i=1}^n |g_i(x)|?$$

Solution:

(a) If there exists x such that $g_i(x) = 0$ for every i, then $\sum_{i=1}^n g_i^2(x) = 0$, which is a global minimum, since the function is non-negative.
(b) The gradient of $f(x) = \sum_{i=1}^n g_i^2(x)$ is $\nabla f(x) = 2 \sum_{i=1}^n g_i(x) \nabla g_i(x)$.
(c) The Hessian is given by $\nabla^2 f(x) = 2 \sum_{i=1}^n (g_i(x) \nabla^2 g_i(x)) + \nabla g_i(x) \nabla g_i(x)^T)$, where $\nabla^2 g_i(x)$ is the Hessian of $g_i(x)$.
(d) Although the global minimum remains as x where $g_i(x) = 0, \forall i = 1, \ldots, n$, the proposed function is not differentiable there at the optimal solution point.

Exercise 2.10

Let $f_i(x) : R^n \to R, 1 \le i \le m$, be a set of real m functions. For some given real numbers c_i, $1 \le i \le m$, define the following function:

$$g(x) = \sum_{i=1}^m (f_i(x) - c_i)^2,$$

and consider the following optimization problem:

$$\min_{x \in R^n} g(x).$$

(a) What are the gradient and Newton's direction of the function $g(x)$ at some given point x?

(b) A suggested iterative method for solving for the value minimizing $g(x)$ is

$$x_{k+1} = \min_{x \in R^n} \sum_{i=1}^{m} \left(f_i(x_k) + \nabla f_i(x_k)^T (x - x_k) - c_i \right)^2.$$

Is this the steepest descent method in disguise? If not, what is the idea behind it?

(c) Show that finding x_{k+1} is a solution to the quadratic optimization of the shape $arg\max_x x^T Q x - 2b^T x$. What are Q and b?

Solution:

(a) Let us first compute the gradient of each summand $(f_i(x) - c_i)^2$. Using the chain rule of derivatives, we have $\nabla (f_i(x) - c_i)^2 = 2(f_i(x) - c_i)\nabla f_i(x)$, and therefore

$$\nabla g(x) = 2 \sum_{i=1}^{m} (f_i(x) - c_i)\nabla f_i(x).$$

To find the Newton's direction of $g(x)$, we first need to compute its Hessian. Using the above equation, we have

$$\nabla^2 g(x) = 2 \sum_{i=1}^{m} \nabla f_i(x)^T \nabla f_i(x) + 2 \sum_{i=1}^{m} (f_i(x) - c_i)\nabla^2 f_i(x),$$

and Newton's direction is given by $-[\nabla^2 g(x)]^{-1}\nabla g(x)$.

(b) This method relies on Taylor's expansion (or approximation) of $f_i(x)$. Therefore, we use the approximation of $f_i(x)$ instead of the function itself, which can be solved using a standard least squares method.

(c) We can easily see that the objective from item (b) is a quadratic function of x, and therefore minimization can be achieved using the least squares method. More specifically, the gradient of the objective is

$$2 \sum_{i=1}^{m} \nabla f_i(x_k) \left(f_i(x_k) + \nabla f_i(x_k)^T (x - x_k) - c_i \right),$$

which vanishes at

$$x = x_k - \left(\sum_{i=1}^{m} \nabla f_i(x_k) \nabla f_i(x_k)^T \right)^{-1} \sum_{i=1}^{m} \nabla f_i(x_k) \left(f_i(x_k) - c_i \right).$$

Thus, this is the same as minimizing a quadratic function of the matrix Q defined as

$$Q = \sum_{i=1}^{m} \nabla f_i(x_k) \nabla f_i(x_k)^T,$$

and the vector b equals

$$b = \sum_{i=1}^{m} \nabla f_i(x_k) \left(f_i(x_k) - c_i \right).$$

Exercise 2.11

Let $X_i \sim Pois(\lambda_i)$ and $Y_i \sim Pois(\beta \lambda_i)$, $1 \leq i \leq n$, be $2n$ independent random variables that follow Poisson distributions, where $\lambda_i > 0, 1 \leq i \leq n$, and $\beta > 0$ are $n+1$ parameters to be estimated.[d]

1. What is the (complete) likelihood function?
2. Prove that the MLEs are

$$\hat{\beta} = \frac{\sum_{j=1}^{n} Y_i}{\sum_{j=1}^{n} X_i} \quad \text{and} \quad \hat{\lambda}_i = \frac{X_i + Y_i}{\hat{\beta} + 1}, \quad 1 \leq i \leq n.$$

3. Suppose that X_1 is a missing datum. Show that the MLEs for λ_i, $1 \leq i \leq n$, and β, based on the incomplete data, solve the following (nonlinear) equations:

$$\hat{\beta} = \frac{\sum_{i=1}^{n} Y_i}{\sum_{i=1}^{n} \hat{\lambda}_i}, \quad \hat{\lambda}_1 = \frac{Y_1}{\hat{\beta}}, \quad \text{and} \quad \hat{\lambda}_i = \frac{X_i + Y_i}{\hat{\beta} + 1}, \quad 2 \leq i \leq n. \quad (2.5)$$

---

[d]This exercise appears in [11, p. 359].

4. Design the resulting EM algorithm (for the incomplete data). In particular, do the following:

   (a) The expectation part: State the expected log likelihood function.

   (b) The maximization part: Show that if the current approximations for the MLEs are $\hat{\beta}^0$ and $\hat{\lambda}_i^0$, $1 \le i \le n$, then the next ones are

$$\hat{\beta}^1 = \frac{\sum_{i=1}^n Y_i}{\hat{\lambda}_1^0 + \sum_{i=2}^n X_i}, \quad \hat{\lambda}_1^1 = \frac{\hat{\lambda}_1^0 + Y_1}{\hat{\beta}^1 + 1}, \quad \text{and}$$

$$\hat{\lambda}_i^1 = \frac{X_i + Y_i}{\hat{\beta}^1 + 1}, \quad 2 \le i \le n. \tag{2.6}$$

Note that $\hat{\lambda}_1^0$ replaces the missing datum of X_1.

5. Show that the fixed point of the iterative procedure stated in (2.6) solves the equations in (2.5).

Solution:

1. The complete likelihood function is given by

$$L(X_1, \ldots, X_n, Y_1, \ldots, Y_n; \beta, \lambda_1, \ldots, \lambda_n)$$

$$= \prod_{i=1}^n \frac{e^{-\lambda_i} \lambda_i^{X_i} e^{-\beta \lambda_i} (\beta \lambda_i)^{Y_i}}{X_i! Y_i!} = e^{-(1+\beta) \sum_{i=1}^n \lambda_i} \cdot \beta^{\sum_{i=1}^n Y_i} \prod_{i=1}^n \frac{\lambda_i^{X_i + Y_i}}{X_i! Y_i!}$$

$$\tag{2.7}$$

2. The log-likelihood function is obtained by:

$$\log(L(X_1, \ldots, X_n, Y_1, \ldots, Y_n; \beta, \lambda_1, \ldots, \lambda_n))$$

$$= -(1 + \beta) \sum_{i=1}^n \lambda_i + \log \beta \sum_{i=1}^n Y_i + \sum_{i=1}^n (X_i + Y_i) \log \lambda_i$$

$$- \sum_{i=1}^n \log(X_i! Y_i!).$$

Taking the derivative with respect to β yields

$$-\sum_{i=1}^n \lambda_i + \frac{\sum_{i=1}^n Y_i}{\beta}. \tag{2.8}$$

Setting this derivative to zero yields

$$\hat{\beta} = \frac{\sum_{i=1}^{n} Y_i}{\sum_{i=1}^{n} \hat{\lambda}_i}.$$

Similarly, the derivative with respect to λ_i equals

$$-(1+\beta) + \frac{X_i + Y_i}{\lambda_i}, \quad 1 \le i \le n.$$

Setting this derivative to zero implies

$$\hat{\lambda}_i = \frac{X_i + Y_i}{1 + \hat{\beta}}, \quad 1 \le i \le n. \tag{2.9}$$

All that remains is to solve the 2×2 system of equations (2.8) and (2.9). Specifically,

$$\hat{\beta} = \frac{\sum_{i=1}^{n} Y_i}{\sum_{i=1}^{n} X_i} \quad \text{and} \quad \hat{\lambda}_i = (X_i + Y_i)\frac{\sum_{j=1}^{n} X_j}{\sum_{j=1}^{n}(X_j + Y_j)}, \quad 1 \le i \le n. \tag{2.10}$$

3. For the incomplete data, we can infer from (2.7) that the log likelihood (up to an additive term that is not a function of the parameters) equals

$$\log L(X_2, \ldots, X_n, Y_1, \ldots, Y_n; \beta, \lambda_1, \ldots, \lambda_n)$$
$$= -(1+\beta) \sum_{i=2}^{n} \lambda_i - \beta\lambda_1 + \log\beta \sum_{i=1}^{n} X_i + \sum_{i=2}^{n}(X_i + Y_i)\log\lambda_i$$
$$+ Y_1 \log \lambda_1. \tag{2.11}$$

Similarly, we can get the MLEs for the incomplete data:

$$\hat{\beta} = \frac{\sum_{i=1}^{n} Y_i}{\sum_{i=1}^{n} \hat{\lambda}_i},$$

$$\hat{\lambda}_1 = \frac{Y_1}{\hat{\beta}} \implies Y_1 = \hat{\beta}\hat{\lambda}_1,$$

$$\hat{\lambda}_i = \frac{X_i + Y_i}{1 + \beta} \implies X_i + Y_i = \hat{\lambda}_i(1 + \hat{\beta}), \quad 2 \le i \le n.$$

These three equations need to be solved. An iterative way to solve these equations is to use the EM algorithm, as outlined below.

4. (a) For the expected log-likelihood and considering the missing X_1, we get an expression similar to (2.11) except X_1 is replaced by its current expected value, namely by $\hat{\lambda}_1^0$.

   (b) Next, we look for resulting equations by taking derivatives and setting them to zero.

   For $2 \leq i \leq n$, equations (2.8) and (2.9) are as before, yielding

   $$\hat{\beta}^1 = \frac{\sum_{i=1}^n Y_i}{\sum_{i=1}^n \hat{\lambda}_i^1}, \qquad (2.12)$$

   $$\hat{\lambda}_i^1 = \frac{X_i + Y_i}{1 + \hat{\beta}^1}, \quad 2 \leq i \leq n. \qquad (2.13)$$

   All that remains is to solve the case for $i = 1$. The derivative of the incomplete likelihood function with respect of λ_1 equals

   $$\frac{\hat{\lambda}_1^0}{\lambda_1} - (1 + \beta) + \frac{Y_1}{\lambda_1}.$$

   Setting this derivative and solving for λ_1 implies that

   $$\hat{\lambda}_1^1 = \frac{\hat{\lambda}_1^0 + Y_1}{1 + \hat{\beta}^1}. \qquad (2.14)$$

   Inspecting equations (2.12), (2.13), and (2.14) indicates that all that is missing is a closed form expression for $\hat{\beta}^1$. This can be obtained by plugging into (2.12) the values stated in (2.13) and in (2.14)

   $$\hat{\beta}^1 = \frac{(1 + \hat{\beta}^1) \sum_{i=1}^n Y_i}{\sum_{i=2}^n X_i + \sum_{i=1}^n Y_i + \hat{\lambda}_1^0},$$

   implying that

   $$\hat{\beta}^1 = \frac{\sum_{i=1}^n Y_i}{\hat{\lambda}_1^0 + \sum_{i=2}^n X_i}.$$

5. Note that the first result obtained in the last maximization was

   $$\hat{\beta}^1 = \frac{\sum_{i=1}^n Y_i}{\sum_{i=1}^n \hat{\lambda}_i^1}$$

   as requires. Next, multiplying both sides of equation (2.14) by $1 + \hat{\beta}^1$ gives us the third equation. Finally, Note that at the end of the process $\hat{\lambda}_1^1 = \hat{\lambda}_1^0$ and $\hat{\beta}^1 = \hat{\beta}^0$, and therefore $\hat{\lambda}_1^1 + \hat{\lambda}_1^1 \hat{\beta}^0 = \hat{\lambda}_1^0 + Y_1$, which implies that $Y_1 = \hat{\lambda}_1 \hat{\beta}^0$, as required.

9.3   Chapter 3

Exercise 3.1

The following problem appears in [1]. First, we define a queueing loss system. The arrival process and the service process are as defined in the queueing examples previously discussed, except now a customer that finds a busy server leaves the system forever. In the case where the arrival rate equals λ and the service rate equals μ, the customers find server idle with a probability of $\mu/(\lambda+\mu)$. Second, assume n such servers exist, where server i serves with a rate of μ_i, $1 \leq i \leq n$. Third, assume a common arrival rate of λ. The question here is how to split the arrival stream among the n servers so as to maximize the expected number of successes per unit of time. Note that if x_i is the arrival rate to server i, then

$$x_i \cdot \frac{\mu_i}{x_i + \mu_i}$$

is the successes rate due to server i, $1 \leq i \leq n$. One wishes to maximize the sum of these values.

1. State formally the decision problem described above as a constrained optimization problem with one equality constraint.
2. Show that the objective function is concave in any of the n decision variables.
3. Argue why in this model all servers are open, namely at the optimal solution $x_i > 0$, $1 \leq i \leq n$.
4. Use the condition of equal derivatives in order to derive the optimal splitting of arrival rate.
5. What is the value of the optimal objective function?
6. What is the derivative of the value stated at the previous item with respect to λ?

Solution:

1.

$$\max_{x_i, 1 \leq i \leq n} \sum_{i=1}^{n} x_i \frac{\mu_i}{x_i + \mu_i},$$

$$\text{s.t.} \ \sum_{i=1}^{n} x_i = \lambda,$$

$$x_i \geq 0.$$

2. The derivative with respect to x_i of the objective function with respect to x_i equals

$$\frac{\mu_i^2}{(x_i + \mu_i)^2}, \quad 1 \le i \le n. \tag{3.15}$$

The second derivative of the objective function with respect to x_i equals

$$-2 \cdot \frac{\mu_i^2}{(x_i + \mu_i)^3}, \quad 1 \le i \le n.$$

Since we have a separable function, the mixed derivatives are zero. Hence, the Hessian matrix is negative, implying that the objective function is concave.

3. Joining a queue for a closed server leads to a probability one of receiving service, which is better than joining the queue of any open server.

4. Denote by α the common derivative. Then, from (3.15), we get that

$$x_i = \frac{\mu_i}{\sqrt{\alpha}} - \mu_i, \quad 1 \le i \le n.$$

Plugging this in the equality constraint leads to an equation in $\sqrt{\alpha}$. Solving this equation yields

$$\alpha = \left(\frac{\sum_{i=1}^n \mu_i}{\lambda + \sum_{i=1}^n \mu_i} \right)^2. \tag{3.16}$$

Therefore,

$$x_i = \mu_i \left(\frac{\lambda + \sum_{j=1}^n \mu_j}{\sum_{j=1}^n \mu_j} - 1 \right) = \mu_i \frac{\lambda}{\sum_{j=1}^n \mu_j}, \quad 1 \le i \le n.$$

5. Plugging this into the objective function, and we get that the optimal value equals

$$\lambda \frac{\sum_{i=1}^n \mu_i}{\lambda + \sum_{i=1}^n \mu_i}.$$

6. The derivative of the above optimal objective value is easily seen to be equal to

$$\left(\frac{\sum_{i=1}^n \mu_i}{\lambda + \sum_{i=1}^n \mu_i} \right)^2,$$

which in fact, as can be seen from (3.16), is equal to common derivative. As we will see in the sensitivity analysis section, this is not a coincidence.

9.4 Chapter 4

Exercise 4.1

Solve the following constrained optimization problem

$$\min_{x \in R^n} \frac{1}{2} \sum_{i=1}^{n} x_i^2,$$

$$\text{s.t. } \sum_{i=1}^{n} x_i = c.$$

From the solution, deduce that

$$\frac{1}{n} \sum_{i=1}^{n} x_i^2 \geq \left(\frac{1}{n} \sum_{i=1}^{n} x_i \right)^2,$$

with equality if and only if $x_1 = x_2 = \cdots = x_n$.

Solution:

Note that the Lagrangian function is given by

$$L(x_1, \ldots, x_n, \lambda) = \frac{1}{2} \|x\|_2^2 - \lambda \left(\sum_{i=1}^{n} x_i - c \right).$$

By Theorem 4.1, the optimal solution x^* satisfies the first-order conditions (FOC):

$$\nabla f(x^*) - \lambda \nabla g(x^*) = x^* - \lambda \cdot \underline{1} = \underline{0} \Rightarrow x^* = \lambda \cdot \underline{1}.$$

This implies $x_i^* \lambda$, $1 \leq i \leq n$. Substituting this into the constraint yields:

$$\sum_{i=1}^{n} \lambda = c \Rightarrow n \cdot \lambda = c \Rightarrow \lambda = \frac{c}{n} \Rightarrow x^* = \frac{c}{n} \cdot \underline{1}.$$

An alternative proof:
Using the constraint and by the Cauchy–Schwarz inequality (see (1.7)),

$$c^2 = \left(\sum_{i=1}^n x_i\right)^2 = \left(\sum_{i=1}^n x_i \cdot 1\right)^2 = (x^T \underline{1})^2 \leq \|x\|_2^2 \cdot \|\underline{1}\|_2^2 = \|x\|_2^2 \cdot n.$$

Hence, $\|x\|_2^2 \geq \frac{c^2}{n}$. Note that if we take $x_i^* = \frac{c}{n} \cdot \underline{1}_n$ for $1 \leq i \leq n$, this inequality is binding. Therefore,

$$\|x^*\|_2^2 = \sum_{i=1}^n (x_i^*)^2 = \sum_{i=1}^n \left(\frac{c}{n}\right)^2 = n \cdot \left(\frac{c}{n}\right)^2 = \frac{c^2}{n}.$$

Thus, the inequality reaches the lower bound and is therefore the minimum point.

Next, observe that for a given $\sum_{i=1}^n x_i = c$,

$$\frac{1}{n} \sum_{i=1}^n x_i^2 \geq \frac{1}{n} \sum_{i=1}^n (x_i^*)^2 = \frac{1}{n} \sum_{i=1}^n \left(\frac{c}{n}\right)^2 = \frac{c^2}{n^2} = \left(\frac{1}{n} \sum_{i=1}^n x_i\right)^2,$$

with equality if and only if $x_i = x_i^* = \frac{c}{n}$ for $1 \leq i \leq n$.

Exercise 4.2

Consider the m constraint optimization problem

$$\min_x f(x),$$

$$\text{s.t. } g_i(x) = 0, \quad 1 \leq i \leq m,$$

and the one-constraint optimization problem

$$\min_x f(x),$$

$$\text{s.t. } \sum_{i=1}^m g_i^2(x) = 0.$$

1. Explain why the two problems are equivalent.
2. Explain why the Lagrange multiplier technique will not work for the latter problem.

Solution:

1. Note that the second constraint $\sum_{i=1}^{m} g_i^2(x) = 0$ is actually the condition $\|g(x)\|_2^2 = 0$. By the non-negativity property of norms, this implies that $g(x) = \underline{0}$, i.e., $g_i(x) = 0$ for $1 \leq i \leq m$.

2. Denote $\sum_{i=1}^{m} g_i^2(x)$ by $h(x)$. Then, the second and equivalent version of the problem is

$$\min_x f(x),$$

$$h(x) = 0.$$

Note that since for any feasible point

$$\nabla h(x) = 2 \sum_{i=1}^{m} g(x_i) \nabla g_i(x) = 2 \sum_{i=1}^{m} 0 \nabla g_i(x) = \underline{0},$$

it follows that the regularity conditions are not met for any feasible point, making the Lagrange multiplier technique inapplicable.

Exercise 4.3

Assume that a symmetric matrix $A \in R^{n \times n}$ has n orthonormal eigenvectors $w_1, \ldots, w_n$, arranged by columns in a matrix W, with corresponding eigenvalues $\lambda_1 \leq \cdots \leq \lambda_n$. Let D be a diagonal matrix, with all the above eigenvalues in its main diagonal. Show that[e]:

(a) $W^T W = I$.
(b) $\max_{\|x\|_2 = 1} x^T A x = \max_{\|x\|_2 = 1} x^T D x = \lambda_n$.
(c) State the corresponding result when "min" is replaced by "max".
(d) Assume now that $\lambda_1 > 0$. Show that

$$\frac{1}{2} = \arg \min_{0 \leq \alpha \leq 1} \frac{\frac{1}{\alpha \lambda_1 + (1-\alpha)\lambda_n}}{\alpha \frac{1}{\lambda_1} + (1-\alpha)\frac{1}{\lambda_n}}. \tag{4.17}$$

Note that, in fact, one maximizes the reciprocal of the function given here. Conclude that the minimum value equals $\frac{4\lambda_1 \lambda_n}{(\lambda_1 + \lambda_n)^2}$, which is the square of the ratio between the geometric mean and the algebraic mean between λ_1 and λ_n.

---

[e]Note that in contrast to Theorem 4.2 above, existence is assumed here.

Solution:

(a) Since the columns of W are orthonormal, we have $W^T W = I$.

(b) We show first that $\max_{\|x\|_2=1} x^T A x = \lambda_n$. Every vector x can be written as a linear combination $x = \sum_{i=1}^n \alpha_i w_i$. Orthonormality of w_i, $1 \le i \le n$, implies: $x^T A x = \sum \alpha_i^2 \lambda_i$. Thus,

$$x^T A x = \sum_{i=1}^n \alpha_i^2 \lambda_i \le \sum_{i=1}^n \alpha_i^2 \lambda_n. \tag{4.18}$$

On the other hand, we have the constraint $\|x\|_2 = 1$, and thus $\|x\|_2^2 = x^T x = 1$. Together with the orthonormality assumption on w_i, $1 \le i \le n$, we have:

$$x^T x = \left(\sum_{i=1}^n \alpha_i w_i \right)^T \left(\sum_{i=1}^n \alpha_i w_i \right) \sum_{i=1}^n \alpha_i^2 \|w_i\|_2^2 = \sum_{i=1}^n \alpha_i^2 = 1.$$

By (4.18), we conclude that $x^T A x \le \lambda_n$. This bound can be reached by the point w_n (i.e., all α_i, $1 \le i \le n$, equal zero except the last one, which equals 1). From the fact that $\max_{\|x\|=1} x^T D x = \lambda_n$, it can be easily seen that $x^T D x = \sum_{i=1}^n \lambda_i x_i^2 \le \lambda_n \sum_{i=1}^n x_i^2 = \lambda_n$, where the last equation follows from the constraint $\|x\| = 1$ (and thus $\|x\|^2 = 1$).

(c) Follow the same steps as previously described, this time with λ_1, reversing the direction of the inequality.

(d) Using some algebra, we obtain an equivalent expression for (4.17):

$$\frac{\lambda_1 \lambda_n}{(\alpha \lambda_1 + (1-\alpha)\lambda_n)(\alpha \lambda_n + (1-\alpha)\lambda_1)}.$$

Since only the denominator depends on the variable α, we want to maximize it. Substituting

$$x = \alpha \lambda_1 + (1-\alpha)\lambda_n \tag{4.19}$$

above, we see that the denominator becomes $x(\lambda_1 + \lambda_n - x)$, which is maximized at the point $x = \frac{\lambda_1 + \lambda_n}{2}$. By (4.19), this is equivalent to $\alpha = \frac{1}{2}$. Now, use $\alpha = \frac{1}{2}$ in (4.17) to obtain the minimum value of $\frac{4\lambda_1 \lambda_n}{(\lambda_1 + \lambda_n)^2}$.

Exercise 4.4

This exercise stems from Exercise 3.3.2 but includes an additional condition: The value of service completion is server dependent. Specifically, the value of service granted by server i is denoted by α_i, $1 \leq i \leq n$.[f]

1. What is the objective function now, and what are the constraints?
2. Show that the objective function is concave in any of its variables.
3. Introduce Lagrange multipliers and state the KTT conditions for optimality.
4. Order the servers in descending order of α_i. Show that if in the optimal solution server i is open, then the same is the case with server $i - 1$. In particular, for some i^s, only the first i^s servers are open. Note that $i^s = n$ is possible.
5. How is α_{i^s+1} compared to the Lagrange multiplier of the equality constraint?
6. Assume that i^s is given. What is the optimal division of the arrival rate and the corresponding value of the objective function?
7. State a procedure for finding i^*.

Solution:

1. The probability of receiving service per arrival at server i, given λ_i, is $\frac{\mu_i}{\mu_i + \lambda_i}$. Hence the expected number of successes in case one goes to server i is $\frac{\mu_i}{\mu_i + \lambda_i}\alpha_i$. This value per unit of time becomes $\lambda_i \frac{\mu_i}{\mu_i + \lambda_i}$, which is then summed over i in order to get the objective function. To get the expected reward, we need to multiply this value by α_i. Thus the objective function becomes

$$\sum_{i=1}^{n} \alpha_i x_i \frac{\mu_i}{x_i + \mu_i}.$$

The constraints are

$$\sum_{i=1}^{n} x_i - \lambda = 0$$

and

$$-x_i \leq 0, \quad 1 \leq i \leq n.$$

---

[f]This problem appears in [1].

We will ignore the inequalities of the signs. Since the solution is derived obeying the KKT conditions, which are sufficient for optimality, and whilst ignoring the constraints above (which obey the KKT conditions regardless), this solution is also optimal had these sign inequalities been introduced.

2. It can be easily seen that the second derivative of the objective over each of the λ_i's is negative, while the Hessian is diagonal (since each variable x_i appears in a separate term). Hence, the Hessian is negative, as required.

3. The Lagrangian function is

$$L(x_i, 1 \leq i \leq n, \gamma, \beta_i, 1 \leq i \leq n)$$

$$= \sum_{i=1}^{n} \frac{\alpha_i \mu_i}{x_i + \mu_i} x_i - \gamma \left(\sum_{i=1}^{n} x_i - \lambda \right) - \sum_{i=1}^{n} \beta_i(-x_i).$$

Note that γ is the multiplier associated with the equality constraints, while β_i is the multiplier associated with the non-negativity constraint $-x_i \leq 0$, $1 \leq i \leq n$.

The KKT conditions are

$$\frac{\alpha_i \mu_i^2}{(x_i + \mu_i)^2} - \gamma + \beta_i = 0, \quad 1 \leq i \leq n,$$

$$\sum_{i=1}^{n} x_i - \lambda = 0,$$

$$x_i \geq 0,$$

$$\beta_i x_i = 0, \quad 1 \leq i \leq n,$$

$$\beta_i \geq 0, \quad 1 \leq i \leq n.$$

Note that the final condition is due to the fact that a maximization problem is solved here.

4. Suppose that $x_i = 0$ and $x_{i+1} > 0$. By introducing a small perturbation $x_i = \epsilon$ and $x_{i+1} \leftarrow x_{i+1} - \epsilon$, such that the solution remains feasible, we can show that the objective increases with ϵ in the neighborhood of $\epsilon = 0$. This renders the current solution suboptimal. The objective function in terms of ϵ equals ϵ is $\frac{\mu_i \alpha_i \epsilon}{\epsilon + \mu_i} + \frac{\mu_{i+1} \alpha_{i+1}(x_{i+1} - \epsilon)}{x_{i+1} - \epsilon + \mu_{i+1}}$. Differentiating by ϵ, we obtain $\frac{\mu_i^2 \alpha_i}{(\epsilon + \mu_i)^2} - \frac{\mu_{i+1}^2 \alpha_{i+1}}{(x_{i+1} - \epsilon + \mu_{i+1})^2}$. At the point $\epsilon = 0$, the derivative equals $\alpha_i - \alpha_{i+1} \frac{\mu_{i+1}^2}{(x_{i+1} + \mu_{i+1})^2}$. Since $\alpha_{i+1} < \alpha_i$, this derivative is positive. Therefore, the objective increases with ϵ when it is close enough to zero.

5. From the previous section, we conclude that $x_i > 0$ for every $1 \leq i \leq i^*$ for some i^*. The KKT conditions imply that $\beta_i = 0$, which leads to $\gamma = \frac{\alpha_i \mu_i^2}{(\lambda_i + \mu_i)^2}$ for $1 \leq i \leq i^*$.

6. From previous items we know that $x_{i^*+1} = 0$. Inserting the first KKT condition, we obtain $\alpha_{i^*+1} - \gamma + \beta_i = 0$. Since $\beta_i \geq 0$, we conclude that $\alpha_{i^*+1} \leq \gamma$.

7. For $i = 1, \ldots, i^*$, all terms $\frac{\alpha_i \mu_i^2}{(x_i + \mu_i)^2}$ are identical. We therefore need to solve the set of equations

$$\gamma = \frac{\alpha_i \mu_i^2}{(x_i + \mu_i)^2}, \quad 1 \leq i \leq i^*$$

$$\sum_{i=1}^{i^*} x_i = \lambda.$$

From the first equation we get that

$$x_i = \frac{\sqrt{\alpha_i} \mu_i}{\sqrt{\gamma}} - \mu_i, \quad 1 \leq i \leq i^*,$$

which coupled with the second equation implies that

$$\sqrt{\gamma} = \frac{\sum_{i=1}^{i^*} \sqrt{\alpha_i} \mu_i}{\lambda + \sum_{i=1}^{i^*} \mu_i} \tag{4.20}$$

and hence,

$$x_i = \sqrt{\alpha_i} \mu_i \frac{\lambda + \sum_{j=1}^{i^*} \mu_j}{\sum_{j=1}^{i^s} \sqrt{\alpha_j} \mu_j} - \mu_i, \quad 1 \leq i \leq i^*. \tag{4.21}$$

The optimal objective value is thus

$$\frac{\sum_{i=1}^{i^*} \sqrt{\alpha_i} \mu_i}{\lambda + \sum_{i=1}^{i^*}} \sum_{i=1}^{i^*} \sqrt{\alpha_i}.$$

8. Set $i = 1$, and repeat the following steps until a stopping criterion is met as defined below. Compute the Lagrange multiplier γ using (4.20). If the inequality $\gamma \geq \alpha_{i+1}$ holds, stop. Set the current i as i^*, and compute all x_i's using (4.21), $1 \leq i \leq i^*$. Otherwise, increment $i \leftarrow i+1$ and repeat the above steps.

Exercise 4.5

Consider the following optimization problem:

$$\min_{x_1, x_2, x_3} \; x_1 + x_2 + x_3,$$

$$\text{s.t.} \quad x_1^2 + x_2 = 3,$$

$$x_1 + 3x_2 + 2x_3 = 7.$$

1. Show that this problem is equivalent to the unconstrained optimization problem $\min_{x_1}(x_1^2 + x_1 + 4)/2$. Solve this problem and deduce the optimal solution for the original problem.
2. Write the Lagrangian function and the KKT conditions for the original problem.
3. Using the fact the optimal solution is given, find the corresponding Lagrange multipliers.
4. Suppose that the 3 and the 7 on the right-hand side above are replaced by 3.01 and 6.98, respectively. Give an educated approximation for the new optimal objective function value.

Solution:

1. From the first constraint, we obtain $x_2 = 3 - x_1^2$. Substituting this into the second constraint gives us

$$x_1 + 9 - 3x_1^2 + 2x_3 = 7 \Rightarrow x_3 = \frac{3}{2}x_1^2 - \frac{x_1}{2} - 1.$$

Now we can express the minimization problem $\min_{(x_1, x_2, x_3)} f(x_1, x_2, x_3) = x_1 + x_2 + x_3$ as

$$\min_{x_1} f(x_1) = x_1 + 3 - x_1^2 + \frac{3}{2}x_1^2 - \frac{x_1}{2} - 1 = \min_{x_1} f(x_1) = \frac{x_1^2 + x_1 + 4}{2}.$$

Setting the derivative of $f(x_1)$ with respect to x_1 to zero, $\frac{df(x_1)}{dx_1} = 0$, we find that $x_1^* + \frac{1}{2} = 0 \Rightarrow x_1^* = -\frac{1}{2}$ (note that this is a convex function of x_1, thus x_1^* is a minimizer), which yields $x_2^* = 2.75$ and $x_3^* = -0.375$.

2. The Lagrangian function is

$$L(x_1, x_2, x_3, \lambda_1, \lambda_2) = x_1 + x_2 + x_3 - \lambda_1(x_1^2 + x_2 - 3) - \lambda_2(x_1 + 3x_2 + 2x_3 - 7),$$

and the KKT conditions are:

$$1 - 2\lambda_1 x_1 - \lambda_2 = 0,$$
$$1 - \lambda_1 - 3\lambda_2 = 0,$$
$$1 - 2\lambda_2 = 0,$$
$$x_1^2 + x_2 - 3 = 0,$$
$$x_1 + 3x_2 + 2x_3 - 7 = 0.$$

3. Solving the system of equations

$$\begin{pmatrix} 1 \\ 1 \\ 1 \end{pmatrix} = \lambda_1 \cdot \begin{pmatrix} 2x_1^* \\ 1 \\ 0 \end{pmatrix} + \lambda_2 \cdot \begin{pmatrix} 1 \\ 3 \\ 2 \end{pmatrix} = \lambda_1 \cdot \begin{pmatrix} -1 \\ 1 \\ 0 \end{pmatrix} + \lambda_2 \cdot \begin{pmatrix} 1 \\ 3 \\ 2 \end{pmatrix},$$

   we obtain $\lambda_1 = -0.5$ and $\lambda_2 = 0.5$.

4. If we replace the numbers 3 and 7 on the right-hand side with 3.01 and 6.98, respectively, which is a very small change in their values, we can use the fact that λ_1 and λ_2 are the shadow prices of the corresponding constraints, and the new optimal objective function value is adjusted by $-0.5 \cdot 0.01 + 0.5 \cdot (-0.02) = -0.015$.

Exercise 4.6

Consider the following constrained minimization problem:

$$\min \frac{1}{2}||x||^2,$$

$$\text{s.t.} \quad Ax = b,$$

where $x \in R^n$ and $A \in R^{m \times n}$ is a full-rank matrix with $m < n$.

(a) Solve this problem while stating the KKT FOCs. In particular, find both the optimal solution and the optimal objective function value. Are the FOCs sufficient? Why?

(b) State the dual function as an unconstrained minimization problem.

(c) Solve the just-mentioned optimization problem.

(d) State the dual optimization problem.

(e) Solve the dual optimization problem.

(f) Compare the objective values of both the primal and dual problems. Do they coincide?

Solution:

(a) First,

$$L(x, \lambda) = \frac{1}{2}x^T A x - \lambda^T (Ax - b).$$

Second, the KKT conditions (4.2) are

$$x - A^T \lambda = \underline{0} \in R^n,$$

$$Ax - b = \underline{0} \in R^m.$$

These equations can be merged into $b = AA^T \lambda$, which becomes $\lambda = (AA^T)^{-1}b$. Hence, $x = A^T (AA^T)^{-1}b$. The optimal objective value therefore becomes $\frac{1}{2}x^T x = \frac{1}{2}b^T (AA^T)^{-1}b$. The KKT FOCs are sufficient because the problem is convex.

(b) The dual function $\phi(\lambda) : R^m \to R$ is therefore

$$\phi(\lambda) = \min_x \frac{1}{2}x^T x - \lambda^T (Ax - b).$$

(c) Differentiating the above objective with respect to x, we get that $x - A^T \lambda = 0$, which yields $x = A^T \lambda$. Inserting this into the objective value, we obtain the dual function

$$\frac{1}{2}\lambda^T AA^T \lambda - \lambda^T (AA^T \lambda - b) = -\frac{1}{2}\lambda^T AA^T \lambda + \lambda^T b.$$

(d) The dual problem is

$$\max_\lambda -\frac{1}{2}\lambda^T AA^T \lambda + \lambda^T b. \tag{4.22}$$

(e) Differentiating the above with respect to λ,

$$-AA^T \lambda + b = \underline{0} \to \lambda = (AA^T)^{-1}b.$$

(f) The maximum objective value is then

$$-\frac{1}{2}b^T (AA^T)^{-1}b + b^T (AA^T)^{-1}b = \frac{1}{2}b^T (AA^T)^{-1}b,$$

which equals the minimal primal objective.

Exercise 4.7

Consider the following optimization problem:

$$\max_{x_1,x_2,x_3} x_1x_2 + x_1x_3 + x_2x_3, \quad \text{s.t.} \quad x_1 + x_2 + x_3 = 3.$$

(a) Convert the problem of finding a stationary point into solving a system of equations.
(b) Replace 3 in the right-hand side with b. For any b in the neighborhood of 3, derive $x_i^*(b), 1 \le i \le 3$, and the optimal objective value as a function of b. What is the derivative of this function when $b = 3$?

Solution:

(a) Using $x_3 = 3 - x_2 - x_1$, we can rewrite the objective as

$$\max_{x_1,x_2}\{x_1x_2+(x_1+x_2)(3-x_2-x_1)\} \implies \max_{x_1,x_2}\{3x_1+3x_2-x_1^2-x_2^2-x_1x_2\}.$$

Deriving the gradient and setting it to zero, we obtain the following set of equations:

$$3 - 2x_1 - x_2 = 0, \quad 3 - 2x_2 - x_1 = 0.$$

The obvious solution is $x_1 = x_2 = 1$, which implies that $x_3 = 1$ too. We can also solve this using Lagrange multipliers as follows: We have

$$L(x_1, x_2, x_3, \lambda) = x_1x_2 + x_1x_3 + x_2x_3 - \lambda(x_1 + x_2 + x_3 - 3).$$

Deriving the gradient and setting it to zero, we obtain the following set of equations:

$$x_2 + x_3 - \lambda = 0,$$
$$x_1 + x_3 - \lambda = 0,$$
$$x_2 + x_1 - \lambda = 0,$$
$$x_1 + x_2 + x_3 = 3,$$

where the last equation is the constraint. The solution of the system of equations above is $x_1 = x_2 = x_3 = 1$, and $\lambda = 2$.

(b) The set of equations is

$$b - 2x_1 - x_2 = 0, \quad b - 2x_2 - x_1 = 0,$$

and $x^*(b) = (\frac{b}{3}, \frac{b}{3}, \frac{b}{3})$. In Lagrangian form, the set of equations is now

$$x_2 + x_3 - \lambda = 0,$$
$$x_1 + x_3 - \lambda = 0,$$
$$x_2 + x_1 - \lambda = 0,$$
$$x_1 + x_2 + x_3 = b.$$

Similar to the previous solution, the solution here is $x_1 = x_2 = x_3 = \frac{b}{3}, \lambda = \frac{2b}{3}$. The optimal objective value as a function of b is $\frac{b^2}{3}$, and its derivative with respect to b is $2b/3$, which coincides with λ, the Lagrange multiplier, in accordance to the envelope theorem, Theorem 4.5.

Exercise 4.8

(a) Solve

$$\min_x x^T Q x, \quad \text{s.t.} \quad Ax = b,$$

where Q is positive, and A is a full-rank matrix, with more columns than rows.

(b) Solve item (a) in the special case where $Q = \begin{bmatrix} 2 & 1 \\ 1 & 2 \end{bmatrix}$, $A = (1, 2)$, and $b = 1$.

(c) Show that for any positive scalar s, the solution to

$$\min_x \left(x^T Q x + s(Ax - b)^T (Ax - b) \right)$$

is

$$\left(\frac{Q}{s} + A^T A \right)^{-1} A^T b.$$

(d) Solve the program in subquestion (c) under the settings of subquestion (b).

(e) Does the solution of (d) in the limit $s \to \infty$ coincide with that of (b)? Prove your answer and explain why.

(f) What is wrong with simply replacing $\frac{Q}{s}$ with zero when we take the limit in the answer of (c)?

Solution:

(a) Define the following Lagrangian function:

$$\min_x x^T Q x - \lambda^T (Ax - b).$$

Take derivative with respect to x and set it to zero

$$2Qx - A^T \lambda = 0 \implies x = \frac{1}{2} Q^{-1} A^T \lambda.$$

Invoking the constraint $Ax = b$, we get that

$$Ax = A(\frac{1}{2} Q^{-1} A^T \lambda) = b \implies \lambda = 2(AQ^{-1}A^T)^{-1}b,$$

which yields the solution

$$x = Q^{-1} A^T (AQ^{-1}A^T)^{-1}b.$$

(b) We simply need to insert the values of Q, A, and b in the expression derived previously for x to obtain $x = (0, 1/2)^T$.

(c) The new objective can be written as a quadratic function of x:

$$x^T Q x + s(Ax - b)^T (Ax - b) = x^T Q x + s(x^T A^T A x - 2b^T Ax + b^T b),$$

whose gradient is given by

$$2Qx + s(2A^T Ax - 2A^T b) = 0 \implies (Qx + sA^T Ax)$$

$$= sA^T b \implies x = (Q + sA^T A)^{-1} sA^T b,$$

which is equivalent to

$$x = \left(\frac{Q}{s} + A^T A\right)^{-1} A^T b.$$

(d) Inserting the values (given in this subquestion) in the solution obtained in subquestion (c), we get that

$$x = \left(\frac{1}{s} \begin{pmatrix} 2 & 1 \\ 1 & 2 \end{pmatrix} + \begin{pmatrix} 1 \\ 2 \end{pmatrix} \begin{pmatrix} 1 & 2 \end{pmatrix}\right)^{-1} \begin{pmatrix} 1 \\ 2 \end{pmatrix} 1$$

To obtain a closed form expression, we can invoke the Sherman–Morrison formula. See Section rank-one. Using $\begin{pmatrix} 2 & 1 \\ 1 & 2 \end{pmatrix}^{-1} =$

$\begin{pmatrix} 2/3 & -1/3 \\ -1/3 & 2/3 \end{pmatrix}$, we get that

$$x = \left(s \begin{pmatrix} 2/3 & -1/3 \\ -1/3 & 2/3 \end{pmatrix} - \frac{s^2 \begin{pmatrix} 2/3 & -1/3 \\ -1/3 & 2/3 \end{pmatrix} \begin{pmatrix} 1 \\ 2 \end{pmatrix} (1 \; 2) \begin{pmatrix} 2/3 & -1/3 \\ -1/3 & 2/3 \end{pmatrix}}{1 + s (1 \; 2) \begin{pmatrix} 2/3 & -1/3 \\ -1/3 & 2/3 \end{pmatrix} \begin{pmatrix} 1 \\ 2 \end{pmatrix}} \right)$$

$$\times \begin{pmatrix} 1 \\ 2 \end{pmatrix},$$

which reduces to $x = (0, s/(1 + 2s))^T$.

(e) It can be easily seen from item (d) that x equals $(0, 1/2)$ in the limit, which coincides with (b).

(f) Since A has more columns than rows, $A^T A$ is singular. Thus, $Q/s + A^T A$ would not be invertible in the case where Q/s is assumed to be the zero matrix.

Exercise 4.9

Consider the following optimization problem (appears in [3, pp. 212–213]):

$$\min_{x_1, x_2, x_3} \quad 2x_1 + 3x_2 - x_3,$$

$$\text{s.t.} \quad x_1^2 + x_2^2 + x_3^2 = 1,$$

$$x_1^2 + 2x_2^2 + 2x_3^2 = 2.$$

(a) Write the KKT conditions.

(b) Show that these set of equations does not have a solution.

(c) Deduce that the optimal solution is irregular, and hence in the optimal solution, the gradients of the constraints are proportional.

(d) Show that the previous item implies that the optimal solution satisfies $x_1^* = 0$ or $x_2^* = x_3^* = 0$.

(e) Rule out the second option in item (d) and conclude that $x_1^* = 0$.

(f) Solve the new problem stemming from the assumption that $x_1^* = 0$.

(g) Prove that the optimal point is $(0, -3/\sqrt{10}, 1/\sqrt{10})$.

(h) What is the worst point (i.e., where is maximization attained)?

Solution:

(a) The KKT conditions are:

$$(2, 3, -1)^T - \lambda_1 (2x_1, 2x_2, 2x_3)^T - \lambda_2 (2x_1, 4x_2, 4x_3)^T = \underline{0}^T.$$

(b) The above set includes the two equations: $-2\lambda_1 x_1 - 4\lambda_2 x_2 = 3$ and $-2\lambda_1 x_1 - 4\lambda_2 x_2 = -1$. Obviously, no solution exists.

(c) If x^* is optimal and regular, then it must satisfy the KKT conditions. Since those conditions do not hold for any point, as seen previously, we conclude that x^* must be irregular. This means that at this point, the constraints' gradients are not linearly independent. Since only two vectors are involved, lack of linear independence means that they are proportional. Thus, the points $(2x_1^*, 2x_2^*, 2x_3^*)$ and $(2x_1^*, 4x_2^*, 4x_3^*)$ are proportional.

(d) If $x_1^* = 0$, then obviously $(2x_1^*, 2x_2^*, 2x_3^*)$ and $(2x_1^*, 4x_2^*, 4x_3^*)$ are linearly dependent. Otherwise, since the first element of the two vectors are non-zero and equal, it follows that $2x_2^* = 4x_2^*$ and $2x_3^* = 4x_3^*$ must also hold (for linear dependence to be satisfied). Thus, $x_2^* = x_3^* = 0$.

(e) The two constraints cannot hold simultaneously if $x_2^* = x_3^* = 0$ (since the first implies $x_1^2 = 1$, while the second implies that $x_1^2 = 2$), and therefore $x_1^* = 0$ is the only possibility left.

(f) Inserting $x_1^* = 0$, we arrive at a new program with a reduced number of variables:

$$\min_{x_2, x_3} \quad 3x_2 - x_3 \quad \text{s.t.} \quad x_2^2 + x_3^2 = 1.$$

(g) From the constraint, we can single out one of the variables, $x_2 = \pm\sqrt{1 - x_3^2}$, and plug it in the objective $\pm 3\sqrt{1 - x_3^2} - x_3$. We wish to minimize this objective, therefore it is sufficient to look at the objective $-3\sqrt{1 - x_3^2} - x_3$, whose derivative equals $\frac{3x_3}{\sqrt{1 - x_3^2}} - 1$. It can be easily seen that the derivative becomes zero at $x_3 = \frac{1}{\sqrt{10}}$, while the second derivative at this point is positive. Hence, the point $x_1^* = 0, x_2^* = \frac{3}{\sqrt{10}}, x_3^* = \frac{1}{\sqrt{10}}$ is a local minimum.

(h) Note that the two constraints imply that $x_2^2 + x_3^2 = 1$ (subtracting the first constraint from the second), which leads to $x_1 = 0$. Plugging this in the objective, together with $x_2 = \pm\sqrt{1 - x_3^2}$ from previously, we obtain $\pm 3\sqrt{1 - x_3^2} - x_3$. Since we wish to maximize that, it is sufficient to look at $3\sqrt{1 - x_3^2} - x_3$, whose derivative equals $-\frac{3x_3}{\sqrt{1 - x_3^2}} - 1$, which vanishes at $x_3 = -\frac{1}{\sqrt{10}}$. The second derivative at this point is negative, and we therefore conclude that $x_1^* = 0, x_2^* = \frac{3}{\sqrt{10}}, x_3^* = -\frac{1}{\sqrt{10}}$ is a maximum point.

Exercise 4.10

Consider the following (primal) optimization problem:

$$\min_{x_1, x_2} \{x_1^2 + x_2^2 + 2x_1\},$$

$$\text{s.t. } x_1 + x_2 = 0.$$

(a) Prove that the objective function is convex, and the feasible set is convex.
(b) Solve it by utilizing a single Lagrange multiplier.
(c) Show that the two sufficient regulation conditions stated in Section 4.2 hold here.
(d) State the dual function and the dual optimization problem.
(e) Solve the dual problem. In particular, show that the optimal dual solution coincides with the value of the Lagrange multiplier of the primal program, and the optimal values of the primal and the dual program coincide.

Solution:

(a) The program is convex since the objective is quadratic and the constraint is affine.
(b) The Lagrangian function is

$$L(x_1, x_2, x_3, \lambda) = x_1^2 + x_2^2 + 2x_1 - \lambda(x_1 + x_2).$$

Setting the derivatives with respect to x_1, x_2 to zero yields the two equations: $2x_1 + 2 - \lambda = 0$ and $2x_2 - \lambda = 0$. Substituting $\lambda = 2x_2$ from the second equation into the first yields $2x_1 + 2 - 2x_2 = 0$. Together with the constraint, we get that

$$x_1 - x_2 = -1,$$

$$x_1 + x_2 = 0.$$

The solution is $x_1 = -0.5, x_2 = 0.5$. Hence $\lambda = 1$. The optimal objective value equals -0.5.

(c) The Lagrangian function is $x_1^2 + x_2^2 + 2x_1 - \lambda(x_1 + x_2)$, the gradient is $\nabla_{x_1,x_2} \to (2x_1 + 2 - \lambda, 2x_2 - \lambda)$, and the Hessian is $\begin{bmatrix} 2 & 0 \\ 0 & 2 \end{bmatrix}$, which is of course positive.

(d) The dual function is $\phi(\lambda) : R \to R$, where

$$\phi(\lambda) = \min_{x_1,x_2} L(x_1, x_2, \lambda).$$

To derive the dual function we need to solve

$$2x_1 + 2 - \lambda = 0, \quad 2x_2 - \lambda = 0,$$

with respect to x_1 and x_2. The solution is

$$x_1 = 0.5\lambda - 1, \quad x_2 = 0.5\lambda.$$

This implies that the dual function is

$$(0.5\lambda - 1)^2 + (0.5\lambda)^2 + 2(0.5\lambda - 1). \tag{4.23}$$

(e) Maximizing this single-variable quadratic function leads to an optimal value of $\lambda^* = 1$, which coincides with what we got in item (b). Moreover, substituting this value into (4.23) leads to the (expected) objective function value of -0.5.

9.5 Chapter 5

Exercise 5.1

Consider the following optimization problem:

$$\min_{x_1,x_2} x_1^2 + 2x_2^2 + 4x_1x_2, \quad \text{s.t.} \quad x_1 + x_2 = 1, \quad x_1, x_2 \geq 0.$$

Convert the problem of finding a stationary point into solving a system of equations.

Solution:

We first present a very short proof that takes advantage of the simple structure of the objective function, which can equivalently be written as $(x_1 + 2x_2)^2 - 2x_2^2$. Under the equality constraint, the objective function becomes $1 - 2x_2 - x_2^2$. This is a quadratic function with a global maximum at $x_2 = -1$. Thus, within the interval $[0, 1]$ (recall that $0 \leq x_2 \leq 1$), the objective function reaches minimum at $x_2 = 1$. As for the long proof, using Lagrange multipliers technique, we can write it as

$$L(x_1, x_2, \lambda, \mu_1, \mu_2) = x_1^2 + 2x_2^2 + 4x_1 x_2 - \lambda(x_1 + x_2 - 1) - \mu_1 x_1 - \mu_2 x_2,$$

which implies the following KKT conditions:

(i) $2x_1 + 4x_2 + \lambda - \mu_1 = 0,$
(ii) $4x_2 + 4x_1 - \lambda - \mu_2 = 0,$
(iii) $x_1 + x_2 - 1 = 0,$
(iv) $x_1, x_2 \geq 0,$
(v) $\mu_1 x_1 = \mu_2 x_2 = 0.$

Note that the KKT conditions are not sufficient, as the problem is not convex (this can be verified by calculating the Hessian of the objective function). We therefore need to consider all solutions that satisfy the KKT conditions, as well as the regularity conditions, and select among them the solution that yields the smallest objective value. Note that the gradients of the left-hand side of the three constraints are given by $(1,1)^T, (1,0)^T, (0,1)^T$. If all inequality constraints are active, then $x_1 = x_2 = 0$ and $x_1 + x_2 = 0 \neq 1$, which renders the solution unfeasible. Our search then narrows down to points that satisfy only one active inequality constraint at the most. All such points are regular, since any subset of the three vectors is linearly independent. A feasible solution must therefore be regular. If $x_1 = 0$, then by (iii), $x_2 = 1$, resulting in an objective function value of 2. Otherwise, if $x_1 \neq 0$, then by (v), $\mu_1 = 0$. The following conditions then must be met:

(i) $2x_1 + 4x_2 - \lambda = 0,$
(ii) $4x_2 + 4x_1 - \lambda - \mu_2 = 0,$
(iii) $x_1 + x_2 = 1,$
(iv) $\mu_2 x_2 = 0,$

which simplifies (by inserting (iii))

(i) $2 + 2x_2 - \lambda = 0,$
(ii) $4 - \lambda - \mu_2 = 0,$
(iii) $\mu_2 x_2 = 0.$

If $\mu_2 = 0$, then by (ii), $\lambda = 4$, and thus, by equation (i), $x_2 = 1$, and therefore $x_1 = 0$. The objective value is 2. If $\mu_2 \neq 0$, then by (iii), $x_2 = 0$ and $x_1 = 1$, and the objective value is 1. Among all the solutions that satisfy the KKT conditions, $x_1 = 1, x_2 = 0$ yields the minimum value.

Exercise 5.2

Consider the following optimization problem:

$$\min_{x \in R^n} F(x),$$

where

$$F(x) = \min_{i=1}^{m} f_i(x),$$

for some m functions $f_i(x) : R^n \to R$, $1 \le i \le m$.

1. Formulate this problem as a constrained optimization problem.
2. What are the FOCs which are met by an optimal solution?
3. Why are regularity conditions not needed to be assumed in the case where no two of the m functions agree for the same x?

Solution:

1.

$$\min \ y,$$

$$\text{s.t.} \quad f_1(x) - y \le 0,$$

$$f_2(x) - y \le 0,$$

$$\vdots$$

$$f_m(x) - y \le 0.$$

2. The Lagrangian function is

$$L(y, x; \mu_j, 1 \le j \le m) = y - \sum_{j=1}^{m} \mu_j(f_j(x) - y).$$

Taking the derivative with respect to y and setting it to zero yields

$$\sum_{j=1}^{m} \mu_j = -1.$$

Taking the derivative (gradient) with respect to x and setting it to $\underline{0} \in R^n$ yields

$$\sum_{j=1}^{m} \mu_j \nabla f_i(x) = \underline{0} \in R^n.$$

Finally, the complementary slackness conditions and the sign condition are

$$\mu_j(y - f_j(x)) = 0 \quad \text{and} \quad \mu_j \leq 0, \quad 1 \leq j \leq m.$$

3. Unless there is a tie at the optimal solution (y^*, x^*), namely for some pair of i and j, $1 \leq i \neq j \leq m$, $y^* = f_i(x^*) = f_j(x^*)$, there is one and only one binding inequality at the optimal solution, making the fulfillment of the regularity condition trivial.

Exercise 5.3

(a) Show that the best way to average a number of uncorrelated unbiased estimators is obtained by giving each one of them a weight that is proportional to the inverse of its variance.

(b) Show that the corresponding optimal variance is the harmonic mean of the variances involved, divided by the number of estimators.

(c) Conclude that the harmonic mean is smaller than or equal to the arithmetic mean.

(d) Now let us remove the assumption that the estimators are uncorrelated. Thus, there exists some variance-covariance matrix Σ that is not necessarily diagonal. What is the optimal estimator now? What is the corresponding variance? Hint: Convert the new mathematical program so that the objective function is to minimize a sum of squares (while changing the constraint accordingly).

Solution:

(a) Consider the set of unbiased estimators $\{T_i\}_{i=1,\ldots,n}$ with variances σ_i^2, $1 \leq i \leq n$. Since the T_i's are uncorrelated, the variance of the weighted average is $\text{Var}(\sum_{i=1}^n w_i T_i) = \sum_{i=1}^n w_i^2 \sigma_i^2$, where $\sum_{i=1}^n w_i = 1$. The optimization problem becomes:

$$\min_{w_1,\ldots,w_n} \sum_{i=1}^n w_i^2 \sigma_i^2, \quad \text{s.t.} \quad \sum_{i=1}^n w_i = 1, \quad w_i \geq 0.$$

Ignore for a while the non-negativity constraints. Denote by λ the Lagrange multiplier of the single equality constraint. Then, the Lagrangian function is

$$L(x_1, \ldots, x_n, \lambda) = \sum_{I=1}^n w_i \sigma_i^2 - \lambda \left(\sum_{i=1}^n w_i - 1 \right).$$

The KKT conditions are

$$2w_i\sigma_i^2 - \lambda = 1, \quad 1 \leq i \leq n,$$

and

$$\sum_{i=1}^{n} w_i = 1.$$

By the first condition, we get that $w_i = \frac{\lambda}{2\sigma_i^2}$, and therefore the weights are proportional to $1/\sigma_i^2$. Since the decision variables need to sum up to 1, the optimal solution is

$$w_i = \frac{\frac{1}{\sigma_i^2}}{\sum_{j=1}^{n} \frac{1}{\sigma_j^2}}, \quad 1, \ldots, n.$$

Since we face a convex optimization problem, this is the global optimal solution.

Finally, since this solution meets the constraints $w_i \geq 0$, $1 \leq i \leq n$, it is also optimal when these set of constraints is considered.

(b) Substituting our result in the objective function, we get that

$$\sum_{i=1}^{n} w_i^2 \sigma_i^2 = \sum_{i=1}^{n} \left(\frac{1}{\sigma_i^2} \frac{1}{\sum_{j=1}^{n} \frac{1}{\sigma_j^2}} \right)^2 \sigma_i^2 = \left(\frac{1}{\sum_{i=1}^{n} \frac{1}{\sigma_i^2}} \right)^2 \sum_{i=1}^{n} \frac{1}{\sigma_i^2},$$

which simplifies to

$$\frac{1}{\sum_{i=1}^{n} \frac{1}{\sigma_i^2}},$$

which is the harmonic mean of the variances divided by n.

(c) The objective value we got is of course smaller than the objective value for $w_i = \frac{1}{n}$, $1 \leq i \leq n$, (the arithmetic mean) because the former is optimal. For this set of weights, the objective function value equals

$$\frac{1}{n^2} \sum_{i=1}^{n} \sigma_i^2,$$

which is the nth fraction of the arithmetic mean of the individual variances.

(d) The new problem can be formulated as

$$\min_{w \in R^n} w^T \Sigma w, \quad \text{s.t.} \quad \underline{1}^T w = 1, \quad w \geq \underline{0}.$$

For simplicity, we allow w to have negative entries. Since Σ is a positive, $\Sigma = U^T \Lambda U$, with diagonal Λ (containing the eigenvalues of Σ, which are positive) and an orthonormal U (containing the corresponding eignvectors). See, e.g., [14, p. 141]. The problem thus becomes

$$\min_{w \in R^n} w^T U^T \Lambda U w, \quad \text{s.t.} \quad \underline{1}^T w = 1.$$

Denote Uw by z, which due to orthonormality of U yields $w = U^T z$ as $U^{-1} = U^T$. Then, the problem we face is equivalent to

$$\min_{z \in R^n} z^T \Lambda z, \quad \text{s.t.} \quad \underline{1}^T U^T z = 1.$$

Solving the latter problem with the Lagrange multiplier technique, we get that

$$L(z, \mu) = z^T \Lambda z - \lambda(\underline{1}^T U^T z - 1).$$

Then,

$$\nabla_z L = 2\Lambda z - \lambda U \underline{1} = \underline{0}, \quad \nabla_\lambda L = -\underline{1}^T U^T z + 1 = 0.$$

Solving the former equation, we get that $z = \frac{1}{2}\mu\Lambda^{-1}U\underline{1}$. Substituting this into the latter, we get that $\lambda = \frac{2}{\underline{1}^T U^T \Lambda^{-1} U \underline{1}} = \frac{2}{\underline{1}^T \Sigma^{-1} \underline{1}}$. The optimal solution becomes $w = U^T z = U^T \frac{1}{2} \frac{2}{\underline{1}^T \Sigma^{-1} \underline{1}} \Lambda^{-1} U^T \underline{1} = \frac{U^T \Lambda^{-1} U \underline{1}}{\underline{1}^T \Sigma^{-1} \underline{1}} = \frac{\Sigma^{-1} \underline{1}}{\underline{1}^T \Sigma^{-1} \underline{1}}$. The optimal objective function value is $\frac{\underline{1}^T \Sigma^{-1} \Sigma \Sigma^{-1} \underline{1}}{(\underline{1}^T \Sigma^{-1} \underline{1})^2} = \frac{1}{\underline{1}^T \Sigma^{-1} \underline{1}}$. Note that subquestion (b) above is just a special case of this result.

Exercise 5.4

Consider the following program:

$$\min_{a,b} \int_{x=0}^{1} (x^2 - ax - b)^2 dx, \quad \text{s.t.} \quad a + b = 1.$$

(a) What are the first- and second-order conditions for a local minimum?
(b) Is this a convex program?
(c) What is the optimal solution, the optimal function value, and the value of the Lagrange multiplier?
(d) Suppose that the constraint $a + b = 1$ is replaced with one of the following constraints:

- $a + b = 1.01$,
- $a + b \leq 1$,
- $a + b \geq 1$.

Describe in words, without doing any further analysis, what you can infer from these three cases from the original version of the program.

Solution:

(a) The Lagrangian function is

$$L(a, b, \lambda) = \int_0^1 (x^2 - ax - b)^2 \, dx - \lambda(a + b - 1).$$

For a given x, let $f(a, b) = (x^2 - ax - b)^2$. Then,

$$\nabla f(a, b) = \begin{bmatrix} -2x(x^2 - ax - b) \\ -2(x^2 - ax - b) \end{bmatrix}.$$

Due to the possibility of interchanging the order of derivatives and integrals here, the FOCs are

$$\int_0^1 (-2x^3 + 2ax^2 + 2xb) \, dx - \lambda = 0,$$

and

$$\int_0^1 (-2x^2 + 2ax + 2b) \, dx - \lambda = 0,$$

which, coupled with the equality constraint, are

$$-\frac{1}{2} + \frac{2}{3}a + b - \lambda = 0,$$

$$-\frac{2}{3} + a + 2b - \lambda = 0,$$

$$a + b = 1.$$

The Hessian of $f(a, b)$ is

$$\nabla^2 f(a, b) = \begin{bmatrix} 2x^2 & 2x \\ 2x & 2 \end{bmatrix},$$

which is a positive matrix.

(b) The objective function is convex, as can be deduced from the fact that the above Hessian matrix is positive. Since the constraints are affine, we have here a convex program, a case where we know that the SOCs are automatically satisfied.

(c) As shown above, the equations derived from the first-order conditions are (calculating the integrals of the gradients derived in item (a)):

$$-\frac{1}{2} + \frac{2}{3}a + b - \lambda = 0,$$

$$-\frac{2}{3} + a + 2b - \lambda = 0,$$

$$a + b = 1,$$

where the third equation is the constraint. The solution is $a = \frac{5}{4}, b = -\frac{1}{4}, \lambda = \frac{1}{12}$. Substituting these values for a and b into the objective function, we get an optimal value of 0.0125.

(d)   (i) Denote by α the right-hand side of the equality constraint. We know that the derivative of the optimal objective w.r.t. α equals λ (the Lagrange multiplier, which is $\lambda = \frac{1}{12}$ in our case where $\alpha = 1$). Thus, using first-order Taylor approximation, we get that that the original objective values has to be increased by $\frac{0.1}{12} = 0.0083$.

  (ii) Since the objective increases with α, the program will "choose" a smaller α, which will result in a smaller objective.

  (iii) For the same reasons as in (ii), the inequality is active. But now, as opposed to (ii), the previous optimal solution stays optimal.

9.6   Chapter 6

Exercise 6.1

Prove Theorem 6.1.

Solution:

Consider the following standard linear program with $A \in R^{m \times n}$:

$$\min_{x} c^T x,$$

$$\text{s.t.} \quad Ax = b,$$

$$x \geq \underline{0}$$

Note that $\min c^T x$ is equivalent to $-\max -c^T x$. By rewriting the constraint $Ax = b$ as $Ax \le b$ and $-Ax \le -b$, we obtain the canonical form:

$$-\max_x -c^T x,$$

$$\text{s.t.} \quad Ax \le b,$$

$$-Ax \le -b,$$

$$x \ge \underline{0}.$$

For the other direction, consider the slack variables $s \in R^n$. Then, rewrite $Ax \le b$ as $Ax + s = b$ for $x \ge 0$ and $s \ge 0$. The standard linear program is thus

$$\min_x c^T x,$$

$$\text{s.t.} \quad Ax + s = b,$$

$$x \ge \underline{0}, \quad s \ge \underline{0}.$$

9.7 Chapter 7

Exercise 7.1

Prove Theorem 7.1.

Solution:

1. Consider the two feasible solutions $x, x' \in S = \{x \mid Ax = b, x \ge 0\}$, and the convex combination $\hat{x} = \lambda x + (1 - \lambda)x'$, $0 \le \lambda \le 1$:

$$A\hat{x} = A(\lambda x) + A((1 - \lambda)x') = \lambda Ax + (1 - \lambda)Ax' = \lambda b + (1 - \lambda)b = b.$$

Hence, $\hat{x} \in S$.

2. Consider the two optimal solutions $x^*, x'^* \in O = \{x^* \mid c^T x^* \le c^T x, \forall x \in S\}$, and the convex combination $\hat{x}^* = \lambda x^* + (1 - \lambda)x'^*$, $0 \le \lambda \le 1$:

$$c^T \hat{x}^* = \lambda c^T x^* + (1 - \lambda)c^T x'^* \le \lambda c^T x + (1 - \lambda)c^T x = c^T x, \quad \forall x \in S.$$

Thus, $\hat{x}^* \in O$.

3. Consider the two right-hand side vectors $b, b' \in B = \{b \mid Ax = b, x \geq 0\}$, and the convex combination $\hat{b} = \lambda b + (1 - \lambda)b'$, $0 \leq \lambda \leq 1$:

$$Ax = \lambda Ax + (1 - \lambda)Ax = \lambda b + (1 - \lambda)b' = \hat{b}$$

The optimality condition holds since it does not depend on the right-hand side vector.

4. Consider the two cost vectors $c, c' \in C = \{c \mid c^T x^* \leq c^T x, \forall x \in S\}$, and the convex combination $\hat{c} = \lambda c + (1 - \lambda)c'$, $0 \leq \lambda \leq 1$:

$$c^T \hat{x}^* = \lambda c^T x^* + (1 - \lambda)c^T x'^* \leq \lambda c^T x + (1 - \lambda)c^T x = c^T x, \quad \forall x \in S.$$

Hence, $\hat{c}$ preserves the optimality condition.

Exercise 7.2

Prove Theorem 7.3.

Solution:

Given a matrix $A \in R^{m \times n}$, let B be an $m \times m$ submatrix comprising m independent columns of A. Since the selection order of columns is irrelevant, the number of distinct feasible bases of A is given by the binomial coefficient $\binom{r}{m}$, where r is the rank of A such that $r \leq n$. This count reaches its maximum when A is of full rank, i.e., $r = n$.

Exercise 7.3

Prove the following technical point: Let A be an invertible matrix such that one of its columns is a unit vector.[g] Show that the corresponding column at A^{-1} is also a unit vector.

Solution:

Consider the invertible matrix $A \in R^{n \times n}$, whose jth column is e_j, a unit vector where its jth entry is 1 and all other entries are 0. Without loss of

---

[g]Recall that we define a unit vector in a matrix as a vector whose diagonal entry is one and whose off-diagonal entries are zero.

generality, assume that $j = 1$. Denote by v the first coloumn of A^{-1}.

$$AA^{-1} = I \implies Av = e_1 = \begin{pmatrix} 1 \\ 0 \\ \vdots \end{pmatrix}. \tag{7.24}$$

Now, consider the submatrix $A_{-1} \in R^{(n-1) \times n}$, the matrix A without its first row, and the homogeneous system of equations $A_{-1}v = 0$. Remove the first column of A_{-1}, which consists of zeros. The resulting $(n-1) \times (n-1)$ matrix is a full-rank one. This implies that the last $n-1$ entries of v are zeros too. Since $A_{11} = 1$, by (7.24), $v_1 = 0$ too.

Exercise 7.4

Apply the Simplex method to the following LP:

$$\min_{x_i, 1 \le i \le 6} -x_1 - 2x_2 + x_3 - x_4 - 4x_5 + 2x_6,$$

$$\text{s.t.} \quad x_1 + x_2 + x_3 + x_4 + x_5 + x_6 \le 6,$$

$$2x_1 - x_2 - 3x_3 + x_4 \le 4,$$

$$x_3 + x_4 + 2x_5 + x_6 \le 4,$$

$$x_i \ge 0, \quad 1 \le i \le 6$$

Solution:

First, convert the LP into a standard LP by adding three slack variables x_7, x_8, and x_9. The cost vector is $c = (-1, -2, 1, -1, -4, 2, 0, 0, 0)^T$. The constraint matrix is:

$$A = \begin{pmatrix} 1 & 1 & 1 & 1 & 1 & 1 & 1 & 0 & 0 \\ 2 & -1 & -3 & 1 & 0 & 0 & 0 & 1 & 0 \\ 0 & 0 & 1 & 1 & 2 & 1 & 0 & 0 & 1 \end{pmatrix}.$$

The right-hand side vector is $b = (6, 4, 4)^T$. The non-negativity constraints are $x_i \ge 0$ for $1 \le i \le 9$, $m = 3$, $n = 9$.

First iteration. Initiate $x_B = (x_7, x_8, x_9)$, $x_N = (x_1, \ldots, x_6)$. Thus,

$$B = \begin{pmatrix} 1 & 0 & 0 \\ 0 & 1 & 0 \\ 0 & 0 & 1 \end{pmatrix} \quad \text{and} \quad B^{-1} = \begin{pmatrix} 1 & 0 & 0 \\ 0 & 1 & 0 \\ 0 & 0 & 1 \end{pmatrix},$$

$$N = \begin{pmatrix} 1 & 1 & 1 & 1 & 1 & 1 \\ 2 & -1 & -3 & 1 & 0 & 0 \\ 0 & 0 & 1 & 1 & 2 & 1 \end{pmatrix}.$$

The initial basic solution is $x_N = (x_1, \ldots, x_6) = \underline{0}^T$ and $x_B = B^{-1}b = b = (6, 4, 4)^T$.

$c_B^T = (0, 0, 0)$, $c_N^T = (-1, -2, 1, -1, -4, 2)$.

Consider the following reduced cost vector

$$\bar{c}_N = c_N^T - c_B^T B^{-1} N = (-1, -2, 1, -1, -4, 2).$$

Since the reduced cost has at least one negative entry, this basic solution can be improved. To decide which variable should be in the basis, let us select the one with the most negative reduced cost: x_5. To decide which variable should be removed, conduct the ratio test:

$$\min_{\substack{i \in B \\ (B^{-1}N)_{i5} > 0}} \left\{ \frac{(B^{-1}b)_i}{(B^{-1}N)_{i5}} \right\} = \min \left\{ \frac{6}{1}, \frac{4}{2} \right\} = 2.$$

Therefore, x_9 is removed from the basis.

Second iteration. $x_B = (x_7, x_8, x_5)$, $x_N = (x_1, x_2, x_3, x_4, x_9, x_6)$. Thus,

$$B = \begin{pmatrix} 1 & 0 & 1 \\ 0 & 1 & 0 \\ 0 & 0 & 2 \end{pmatrix} \quad \text{and} \quad B^{-1} = \begin{pmatrix} 1 & 0 & -0.5 \\ 0 & 1 & 0 \\ 0 & 0 & 0.5 \end{pmatrix},$$

$$N = \begin{pmatrix} 1 & 1 & 1 & 1 & 0 & 1 \\ 2 & -1 & -3 & 1 & 0 & 0 \\ 0 & 0 & 1 & 1 & 1 & 1 \end{pmatrix}.$$

Now, $x_N = \underline{0}$ and $x_B = B^{-1}b = (4, 4, 2)^T$.

$c_B^T = (0, 0, -4)$, $c_N^T = (-1, -2, 1, -1, 0, 2)$,
and

$$\bar{c}_N = c_N^T - c_B^T B^{-1} N = (-1, -2, 1, -1, 0, 2) - (0, 0, -2, -2, -2, -2)$$
$$= (-1, -2, 3, 1, 2, 4)$$

Since the reduced cost has at least one negative entry, this basic solution can be improved further. Now x_2 has the most negative reduced cost, and so it is in the basis. For the removed variable:

$$\min_{\substack{i \in B \\ (B^{-1}N)_{i2} > 0}} \left\{ \frac{(B^{-1}b)_i}{(B^{-1}N)_{i2}} \right\} = \min \left\{ \frac{4}{1} \right\} = 4$$

Therefore, the removed variable is x_7.

Third iteration. $x_B = (x_2, x_8, x_5)$, $x_N = (x_1, x_7, x_3, x_4, x_9, x_6)$. Thus,

$$B = \begin{pmatrix} 1 & 0 & 1 \\ -1 & 1 & 0 \\ 0 & 0 & 2 \end{pmatrix} \quad \text{and} \quad B^{-1} = \begin{pmatrix} 1 & 0 & -0.5 \\ 1 & 1 & -0.5 \\ 0 & 0 & 0.5 \end{pmatrix},$$

$$N = \begin{pmatrix} 1 & 0 & 1 & 1 & 0 & 1 \\ 2 & 0 & -3 & 1 & 0 & 0 \\ 0 & 0 & 1 & 1 & 1 & 1 \end{pmatrix}.$$

Now, $x_N = \underline{0}$ and $x_B = B^{-1}b = (4, 8, 2)^T$.
$c_B^T = (-2, 0, -4)$, $c_N^T = (-1, 0, 1, -1, 0, 2)$,
and

$$\bar{c}_N = c_N^T - c_B^T B^{-1} N = (-1, 0, 1, -1, 0, 2) - (-2, -2, -3, -3, -1, -3)$$
$$= (1, 2, 4, 2, 1, 5).$$

We get that all entries of the reduced cost are positive, and so the current basic solution is the optimal one, namely $x_B = (x_2, x_8, x_5) = (4, 8, 2)$ and $x_N = (x_1, x_7, x_3, x_4, x_9, x_6) = \underline{0}$. The value of the objective function is $c_B^{TT} x_B = -16$.

Exercise 7.5

(Caratheodory theorem) Let $x^1, x^2, \ldots, x^k$ be a set of k points in R^n. Let C be the convex hull of these points, namely

$$ C = \left\{ x \in R^n, \ x = \sum_{i=1}^{k} \alpha_i x^i \ \middle| \ \sum_{i=1}^{k} \alpha_i = 1, \alpha_i \geq 0, 1 \leq i \leq k \right\}. $$

In other words, C is the set of all convex combinations of $x^1, x^2, \ldots, x^k$. Show that any $x \in C$ can be expressed as a convex combination of at most $n + 1$ points out of $x^1, x^2, \ldots, x^k$.[h]

Solution:

Proof. Fix some $x \in C$. We are looking for the set of k decision variables α_i, $1 \leq i \leq$ that satisfy

$$ \sum_{j=1}^{k} \alpha_j x_i^j = x_i, \quad 1 \leq i \leq n, $$

$$ \sum_{j=1}^{n} \alpha_j = 1, $$

$$ \alpha_j \geq 0, \quad 1 \leq j \leq k. $$

This set of constraints is the same as that of a standard LP, with an objective that is not relevant to our discussion. This set is known to be feasible. Hence, it has at least one basic solution. Since the number of equality constraints is $n + 1$, it is also the number of basic variables. The rest are non-basic variables and their values in this solution are zero.

Exercise 7.6

Let X be the set defined by $X = \{x | Ax = b, x \geq 0\}$. The non-zero vector d is said to be a *ray* for $x \in X$ if for any non-negative scalar α, $x + \alpha d \in X$.

- Show that d is a ray for $x \in X$ if any only if it is a ray for $y \in X$.
- Show that d is a ray for any $x \in X$ if and only if d satisfies $Ad = 0$ and all its entries are non-negative.

---

[h]Of course, these $n + 1$ points vary with the selected x.

Solution:

1. Consider d as a ray from $x \in X$ to $y \in X$. Then

$$A(y + \lambda d) = Ay + \lambda Ad = b + \lambda Ad = Ax + \lambda Ad = A(x + \lambda d) \in X,$$

since d is a ray. We get the converse by interchanging the roles of x and y.

2. Assume that d is a ray; then $A(x + \lambda d) = b$ for all $x \in X$ with $x \geq 0$.

$$\Rightarrow Ax + \lambda Ad = b \Rightarrow \lambda Ad = 0 \text{ for all } \lambda \geq 0 \Rightarrow Ad = 0.$$

For non-negativity, $x + \lambda d \in X \Rightarrow x + \lambda d \geq 0$. Assume that there is an entry j such that $d_j < 0$. However, we can then choose $\tilde{\lambda} = -(x_j/d_j + \varepsilon)$ for a sufficiently large $\varepsilon > 0$, leading to $x_j + \tilde{\lambda} d_j < 0$, which is a contradiction.

Assume that $Ad = 0$ and $d \geq 0$. Then $A(x + \lambda d) = Ax + \lambda Ad = b$, and $x + \lambda d$ is non-negative for all $\lambda \geq 0$, which implies that $x + \lambda d \in X$.

Exercise 7.7

Draw the following set X:

$$-3x_1 + x_2 \leq -2,$$
$$-x_1 + x_2 \leq 2,$$
$$-x_1 + 2x_2 \leq 8,$$
$$x_2 \geq 2.$$

- What are the extreme points and extreme rays of X?
- Express $(4, 3)$ as a convex combination of the extreme points plus a non-negative combination of the extreme rays.

Solution:

The set is pictured in the Figure 9.1.

1. The extreme points must belong to the feasible set and cannot be expressed as a linear combination of other feasible points. The graph shows three extreme points. Calculating the intersections between the

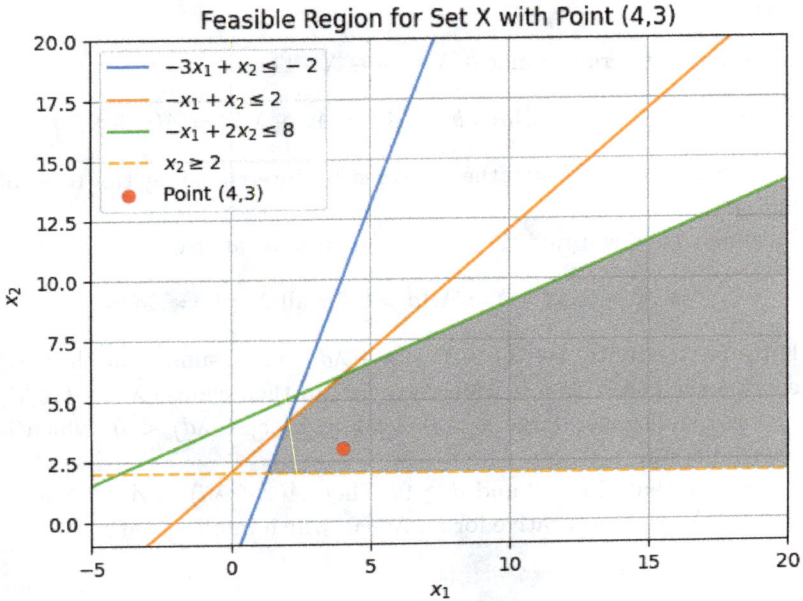

Figure 9.1. Feasible region for set X with point (4,3).

constraints in each one of these three points yields their coordinates: $(2, 4)$, $\left(\frac{4}{3}, 2\right)$, $(4, 6)$.

2. To find the extreme rays, since $x \in R^2$, we need to identify (non-zero) directions that preserve the original inequalities with a zero vector instead of the right-hand side vector b, with at least one inequality being binding: For $-3d_1 + d_2 = 0 \Rightarrow d_2 = 3d_1$, which, combined with $d_2 \geq 0$, violates the second inequality $-d_1 + d_2 \leq 0$. For $-d_1 + d_2 = 0 \Rightarrow d_1 = d_2$: $d_2 \geq 0 \Rightarrow -d_1 + 2d_2 > 0$, which violates the third constraint. For $-d_1 + 2d_2 = 0 \Rightarrow d_1 = 2d_2$, which satisfies all other constraints, and so we take the direction $(d_1, d_2) = (1, 0.5)$ to be the first ray. For $d_2 = 0$, any positive d_1 satisfies all other constraints, so we take the other ray to be $d = (1, 0)$.

3. We need to express the vector $\begin{pmatrix} 4 \\ 3 \end{pmatrix}$ as a linear combination of the three extreme points and the two extreme rays such that

$$\begin{pmatrix} 4 \\ 3 \end{pmatrix} = \lambda_1 \cdot \begin{pmatrix} 2 \\ 4 \end{pmatrix} + \lambda_2 \cdot \begin{pmatrix} \frac{4}{3} \\ 4 \end{pmatrix} + \lambda_3 \cdot \begin{pmatrix} 4 \\ 6 \end{pmatrix} + \mu_1 \cdot \begin{pmatrix} 1 \\ \frac{1}{2} \end{pmatrix} + \mu_2 \cdot \begin{pmatrix} 1 \\ 0 \end{pmatrix},$$

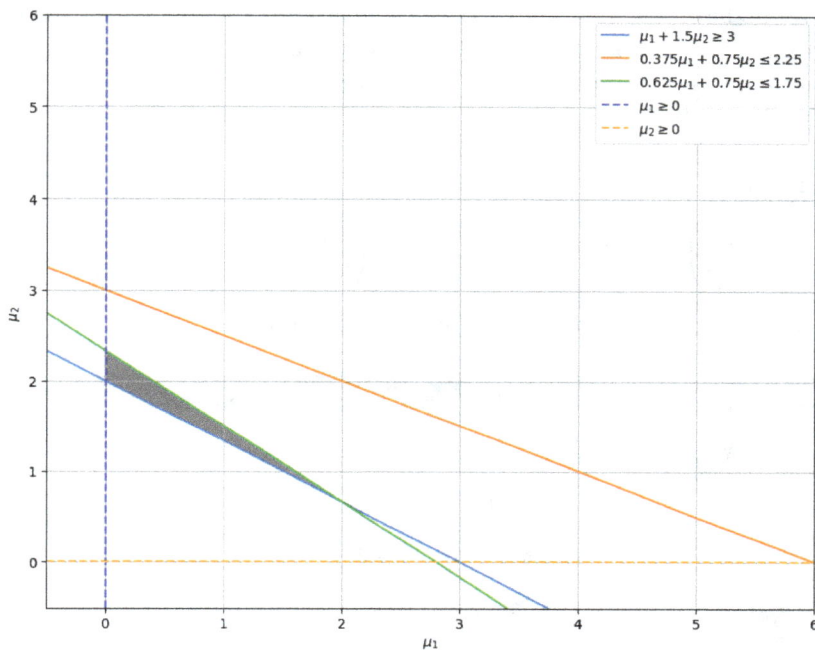

Figure 9.2. Feasible region for μ_2, μ_2.

with $\sum_{i=1}^{3} \lambda_i = 1$, $\lambda_i \geq 0$, $1 \leq i \leq 3$, and $\mu_j \geq 0$, $1 \leq j \leq 2$. This system is solved by reducing the augmented matrix

$$
\begin{pmatrix}
2 & \frac{4}{3} & 4 & 1 & 1 \\
4 & 2 & 6 & \frac{1}{2} & 0 \\
1 & 1 & 1 & 0 & 0
\end{pmatrix}
\cdot
\begin{pmatrix}
\lambda_1 \\
\lambda_2 \\
\lambda_3 \\
\mu_1 \\
\mu_2
\end{pmatrix}
=
\begin{pmatrix}
4 \\
3 \\
1
\end{pmatrix},
$$

which simplifies to:

$$
\begin{pmatrix}
1 & 0 & 0 & -1 & -1.5 \\
0 & 1 & 0 & 0.375 & 0.75 \\
0 & 0 & 1 & 0.625 & 0.75
\end{pmatrix}
\cdot
\begin{pmatrix}
\lambda_1 \\
\lambda_2 \\
\lambda_3 \\
\mu_1 \\
\mu_2
\end{pmatrix}
=
\begin{pmatrix}
-3 \\
2.25 \\
1.75
\end{pmatrix}.
$$

As we have five unknowns and three equations, we have an underdetermined set of equations. We also need to mind the non-negativity constraints. The non-negativity of λ_i, $1 \leq i \leq 3$, implies that $\mu_1 + 1.5\mu_2 \geq 3$, $0.375\mu_1 + 0.75\mu_2 \leq 2.25$, and $0.625\mu_1 + 0.75\mu_2 \leq 1.75$. See Figure 9.2. By choosing $\mu_1 = 2$ and $\mu_2 = \frac{2}{3}$, we find $\lambda_1 = 0$, $\lambda_2 = 1$, and $\lambda_3 = 0$. To summarize:

$$\begin{pmatrix} 4 \\ 3 \end{pmatrix} = 0 \cdot \begin{pmatrix} 2 \\ 4 \end{pmatrix} + 1 \cdot \begin{pmatrix} \frac{4}{3} \\ 2 \end{pmatrix} + 0 \cdot \begin{pmatrix} 4 \\ 6 \end{pmatrix} + 2 \cdot \begin{pmatrix} 1 \\ \frac{1}{2} \end{pmatrix} + \frac{2}{3} \cdot \begin{pmatrix} 1 \\ 0 \end{pmatrix}.$$

It is not claimed that this is the unique solution.

Exercise 7.8

The following LP is given:

$$\max_{x_1, x_2, x_3} 2x_1 + x_2 - x_3,$$

$$\text{s.t.} \quad x_1 + x_2 + 2x_3 \leq 6,$$

$$x_1 + 4x_2 - x_3 \leq 4,$$

$$x_i \geq 0, \quad 1 \leq i \leq 3.$$

- What is the set of extreme points of the feasible set?
- What is the (possibly empty) set of extreme rays?
- Express any feasible point as a convex combination of the extreme points plus a non-negative combination of the extreme rays. Can you now guess the optimal solution?

Solution:

1. To find the extreme points, we need to identify all the basic solutions. This can be done by adding two slack variables x_4 and x_5 to the LP:

$$\max_{x_1, x_2, x_3} 2x_1 + x_2 - x_3$$

subject to

$$x_1 + x_2 + 2x_3 + x_4 = 6,$$

$$x_1 + 4x_2 - x_3 + x_5 = 4,$$

$$x_i \geq 0, \quad 1 \leq i \leq 5.$$

The basic solutions are with the possible set of basic variables: (x_1, x_4), (x_1, x_3), (x_2, x_4), (x_2, x_3), (x_3, x_5), which yield the extreme points (x_1, x_2, x_3): $(4, 0, 0)$, $(\frac{14}{3}, 0, \frac{2}{3})$, $(0, 1, 0)$, $(0, \frac{14}{9}, \frac{20}{9})$, $(0, 0, 3)$, respectively.

2. To find the extreme rays, since $x \in R^3$, we need to identify (non-zero) directions that preserve the original inequalities, but with a zero vector instead of the right-hand side vector b. One of the conditions is $d_1 + d_2 + 2d_3 = 0$, which by the non-negativity constraint is satisfied only by $d_i = 0$, $1 \leq i \leq 3$. Thus, the set of extreme rays is empty.

3. For $y \in X = \{x \in R^3 \mid Ax \leq b, x \geq \underline{0}\}$, which corresponds to the above constraints, we can express y as

$$\begin{pmatrix} y_1 \\ y_2 \\ y_3 \end{pmatrix} = \alpha_1 \cdot \begin{pmatrix} 4 \\ 0 \\ 0 \end{pmatrix} + \alpha_2 \cdot \begin{pmatrix} \frac{14}{3} \\ 0 \\ \frac{2}{3} \end{pmatrix} + \alpha_3 \cdot \begin{pmatrix} 0 \\ 1 \\ 0 \end{pmatrix} + \alpha_4 \cdot \begin{pmatrix} 0 \\ \frac{14}{9} \\ \frac{20}{9} \end{pmatrix} + \alpha_5 \cdot \begin{pmatrix} 0 \\ 0 \\ 3 \end{pmatrix},$$

with $\sum_{i=1}^{5} \alpha_i = 1$. From examining the objective function, it can be easily seen that in this set, the maximum value is reached by taking the coefficient of $(\frac{14}{3}, 0, \frac{2}{3})$ to be 1, i.e., as large as possible, and keeping all the other coefficients at 0. In particular, $(x_1, x_2, x_3) = (\frac{14}{3}, 0, \frac{2}{3})$ is the optimal LP solution.

9.8 Chapter 8

Exercise 8.1

A real function f is called *sub-additive* if for any pair x_1 and x_2, $f(x_1 + x_2) \leq f(x_1) + f(x_2)$. It is called *super-additive* in the case where the reverse inequality holds. Consider the LP

$$\max_{x} c^T x,$$

$$\text{s.t.} \quad Ax = b,$$

$$x \geq \underline{0}.$$

- Show that the value of the objective function is a sub-additive function of the cost coefficient vector c.
- Show that the value of the objective function is a super-additive function of the right-hand side vector b.

Solution:

1. Consider the two LPs:

$$\max_{x_1} c_1^T x_1$$

subject to

$$A x_1 = b, \quad x_1 \geq \underline{0},$$

and

$$\max_{x_2} c_2^T x_2$$

subject to

$$A x_2 = b, \quad x_2 \geq \underline{0}.$$

Now consider $c = c_1 + c_2$ and the LP:

$$\max_{x} c^T x$$

subject to

$$A x = b, \quad x \geq \underline{0}.$$

Let x_1^*, x_2^*, and x^* be the optimal solutions for the above three LPs and in that order. By way of contradiction, assume that $c^T x^* > c_1^T x_1^* + c_2^T x_2^*$. This implies that $c_1^T x^* + c_2^T x^* > c_1^T x_1^* + c_2^T x_2^*$, and hence $c_1^T x^* > c_1 x^*$ and/or $c_2^T x^* > c_2^T x_2^*$. Since $A x^* = A x_1^* = A x_2^*$, x^* is also a feasible point for each of the first two LPs, and hence either x_1^* or x_2^* can be improved, contradicting their optimality.

2. Consider the two LPs:

$$\max_{x_1} \ c^T x_1$$

subject to

$$A x_1 = b_1, \quad x_1 \geq \underline{0},$$

and

$$\max_{x_2} \ c^T x_2$$

subject to

$$A x_2 = b_2, \quad x_2 \geq \underline{0}.$$

Now consider $b = b_1 + b_2$ and the LP:

$$\max_x \ c^T x$$

subject to

$$Ax = b, \quad x \geq \underline{0}.$$

Assume, by way of contradiction, that $c^T x_1^* + c^T x_2^* > c^T x^*$. Note that $Ax^* = Ax_1^* + Ax_2^* = b_1 + b_2 = b$, and therefore $x' = x_1^* + x_2^*$ is a feasible point for the third LP. However, $c^T x' > c^T x^*$ for a feasible point x', which contradicts the optimality of x^*.

Exercise 8.2

For the example given in Section 7.3.5, answer the following questions:

1. What are the shadow (dual) prices for any of the right-hand side entries?

2. For the first two right-hand side entries, determine the ranges in which these entries can vary (when all other coefficients are fixed at their original values) while the optimal basis is maintained.

3. (Cont.) Give an explicit expression for the values of the optimal solutions and for the optimal objective functions in these ranges.

4. What is the optimal solution and what is the value of the objective function when the first entry in the right-hand side is changed from 48 to 30, and to 60?

5. What is the optimal solution and what is the value of the objective function when the second entry in the right-hand side is changed from 20 to 22, and to 18?

6. Suppose that a new variable, say shelves, is suggested. Its price coefficient is 10 and its constraint column is $(2, 1, 0.5, 0)^T$. Should any production of shelves be initialized?

7. Suppose that the right-hand side is changed to $(55, 19, 8, 5)$. Does the optimal basis change? If not, what is the new optimal solution and what is the new value of the objective function?

8. For each of the non-basic cost coefficients find the range in which they can vary (when all other coefficients are fixed at their original values) and the optimal solution is maintained.

9. (Cont.) Give an expression for the optimal solution and for the optimal value of the objective function. Hint: The answer is trivial.

10. For the non-slack variable in the basis with the lowest index, find the range in which its cost coefficient can change when the optimality of the optimal basis is maintained.

Solution:

1. Recall that we obtained $c_B^T = (0, -20, -60, 0)$, $b = (48, 20, 8, 4)$, and

$$
B^{-1} = \begin{pmatrix} 1 & 2 & -8 & 0 \\ 0 & 2 & -4 & 0 \\ 0 & -0.5 & 1.5 & 0 \\ 0 & 0 & 0 & 1 \end{pmatrix}
$$

for the basic variables $x_B = (x_4, x_3, x_1, x_7) = B^{-1}b = (24, 8, 2, 4)$. The shadow price for the ith right-hand side entry is given by: $\frac{\partial f(b)}{\partial b_i} = c_B^T \cdot B_{\cdot i}^{-1}$, which gives us the shadow price vector $(0, -10, -10, 0)$.

2. The range of the ith right-hand side entry that maintains the optimal basis is $b_i \in [b_i^* - \theta_i^-, b_i^* + \theta_i^+]$, where b_i^* is the ith entry of the current right-hand side vector, $\theta_i^+ = \min_{\substack{j \in B, \\ B_{ji}^{-1} < 0}} \left\{ -\frac{(B^{-1}b)_j}{(B_{ji}^{-1})} \right\}$, and

$$
\theta_i^- = \min_{\substack{j \in B, \\ B_{ji}^{-1} > 0}} \left\{ \frac{(B^{-1}b)_j}{(B_{ji}^{-1})} \right\}.
$$

For $i = 1$, all the values of B_{j1}^{-1} are ≥ 0, and therefore θ_1^+ is infinite. Since only the first entry is positive, $\theta_1^- = \min\{24\} = 24$.

For $i = 2$, only the third entry of $B_{\cdot 2}^{-1}$ is negative; then $\theta_2^+ = \min\left\{ \frac{-2}{-0.5} \right\} = 4$. Thus, $\theta_2^- = \min\left\{ \frac{24}{2}, \frac{8}{2} \right\} = 4$.

Therefore, we get that $b_1 \in [24, \infty)$, and $b_2 \in [16, 24]$.

3. The optimal solution is $x_B = B^{-1}b$. Then $x_B' = B^{-1}b^* + B^{-1}\Delta b \cdot e_i = x_B^* + B^{-1}\Delta b \cdot e_i$. For b_1, we obtain

$$
\begin{pmatrix} 24 \\ 8 \\ 2 \\ 4 \end{pmatrix} + \begin{pmatrix} 1 & 2 & -8 & 0 \\ 0 & 2 & -4 & 0 \\ 0 & -0.5 & 1.5 & 0 \\ 0 & 0 & 0 & 1 \end{pmatrix} \cdot \begin{pmatrix} \Delta b_1 \\ 0 \\ 0 \\ 0 \end{pmatrix} = \begin{pmatrix} 24 + \Delta b_1 \\ 8 \\ 2 \\ 4 \end{pmatrix}.
$$

Note that the change in the value of x_B is only in the slack variable, consistent with the zero shadow price of b_1, meaning it does not affect the value of the objective function.

For b_2, we get that

$$
\begin{pmatrix} 24 \\ 8 \\ 2 \\ 4 \end{pmatrix} + \begin{pmatrix} 1 & 2 & -8 & 0 \\ 0 & 2 & -4 & 0 \\ 0 & -0.5 & 1.5 & 0 \\ 0 & 0 & 0 & 1 \end{pmatrix} \cdot \begin{pmatrix} 0 \\ \Delta b_2 \\ 0 \\ 0 \end{pmatrix} = \begin{pmatrix} 24 + 2\Delta b_2 \\ 8 + 2\Delta b_2 \\ 2 - 0.5\Delta b_2 \\ 4 \end{pmatrix}.
$$

The shadow price of b_2 is -10, and therefore the value of the objective function changes by $-10 \cdot \Delta b_2$.

4. When the first entry in the right-hand side vector is changed from 48 to 30, which is within the range found in (2), the change in the optimal solution is only in the value of x_4 (a slack variable) from 24 to 6. As mentioned, the value of the objective function remains the same. For 60, which is also within the range, the value of x_4 changes from 24 to 36, and the objective function does not change.

5. When the second entry in the right-hand side vector changes from 20 to 22, which is within the allowable range, then $x'_B = (28, 12, 1, 4)$, and the objective function changes from -280 to -300. If we change the right-hand side to 18, $x'_B = (20, 4, 3, 4)$, and the value of the objective function increases to -260.

6. As we have seen, a new variable j is added to the basis if and only if $c_j - \sum_{i=1}^{m} c_B^T B_{\cdot i}^{-1} A_{ij} = \sum_{i=1}^{m} \pi_i A_{ij} < 0$. Recall we are dealing with a maximization problem, and therefore $c_j = -10$. Plugging in the given information yields:

$$
-10 - (0, -10, -10, 0) \begin{pmatrix} 2 \\ 1 \\ 0.5 \\ 0 \end{pmatrix} = 5 > 0,
$$

and therefore the basic variables remain the same.

7. Since both changes in the right-hand side entries associated with the slack variables are positive, and as previously mentioned, the amount of increase which preserve feasibility is unlimited, and since b_2 is within its range, all the changes, which sum up to less than 100% (precisely, 25%), we can deduce that the optimal basis does not change. The new optimal

solution is $x'_B = (x_4, x_3, x_1, x_7) = B^{-1}(b^* + \Delta b_1 \cdot e_1 + \Delta b_2 \cdot e_2 + \Delta b_4 \cdot e_4) =$
$\begin{pmatrix} 29 \\ 6 \\ 2.5 \\ 5 \end{pmatrix}$, and the value of the objective function is $c_B^{TT} x'_B = -270$.

8. The only non-basic variable is x_2 with its cost coefficient of -30. The possible change in the cost coefficient that preserves the original optimal solution is the one that makes the reduced cost of x_2 equal to zero:

$$\bar{c}_j = c_j - c_B^T B^{-1} N_{.j} = 0 \Rightarrow c_j = (0, -10, -10, 0) \cdot \begin{pmatrix} 6 \\ 2 \\ 1.5 \\ 4 \end{pmatrix} = -35.$$

9. Since the optimal solution remains the same, the value of the objective function is -280.

10. In the same manner as in (8), we can calculate the minimal cost coefficient for the slack variables as well, obtaining $c_5 = c_6 = -10$.

Exercise 8.3

The following LP is given:

$$\min_x c^T x,$$
$$\text{s.t.} \quad Ax \geq b,$$
$$x \geq \underline{0},$$

with x^* as an optimal solution. Let A_1, A_2, b_1, b_2 be defined as follows:

$$A = \begin{pmatrix} A_1 \\ A_2 \end{pmatrix} \quad \text{and} \quad b = \begin{pmatrix} b_1 \\ b_2 \end{pmatrix}.$$

It is given that $A_1 x^* = b_1$ and $A_2 x^* > b_2$. Show that x^* is also an optimal solution to the following two LPs:

$$\min_x c^T x,$$
$$\text{s.t.} \quad A_1 x \geq b_1,$$
$$x \geq \underline{0},$$

and

$$\min_x c^T x,$$

$$\text{s.t.} \quad A_1 x = b_1,$$

$$x \geq \underline{0}.$$

Solution:

Consider x^* as an optimal solution to the following linear program (LP):

$$\min_x \ c^T x$$

subject to

$$\begin{pmatrix} A_1 \\ A_2 \end{pmatrix} x \leq b, \quad x \geq \underline{0}.$$

It is given that $A_1 x^* = b_1$ and $A_2 x^* < b_2$. Now, consider the LP with a modified constraint set:

$$\min_x \ c^T x$$

subject to

$$A_1 x \leq b_1, \quad x \geq \underline{0}.$$

For feasibility, x^* is obviously feasible under the modified constraints since it was feasible under the stricter constraints. For optimality, assume by way of contradiction that there exists an x' in the feasible region of the modified LP where $cx' \leq cx^*$. If $A_2 x' \geq b_2$, then x' would also be feasible for the original LP, contradicting the optimality of x^*. Therefore, x' must satisfy $A_2 x' > b_2$. Given that for x^*, $A_2 x^* < b_2$ and $A_2 x^* = b_2$, making x^* a feasible point, it follows that the constraints associated with b_2 are not binding. Consequently, all basic variables related to b_1 have slack variables for b_2. Imposing the new set of constraints is effectively the same as increasing all entries of b_2 in the original LP to ∞. The allowable increase θ_i^+ to maintain basis optimality is then

$$\theta_i^+ = \min_{\substack{j \in B, \\ B_{ji}^{-1} < 0}} \left\{ -\frac{(B^{-1}b)_j}{(B_{ji}^{-1})} \right\},$$

since we have slack in these constraints, the minimization is taken over an empty set, indicating that the allowable increase is infinite. Thus, x^* remains optimal for the modified LP as well.

By similar reasoning, x^* is also optimal for the LP with the following constraints:

$$\min_{x} \; c^T x$$

subject to

$$A_1 x = b_1, \quad x \geq \underline{0},$$

which are stricter than the modified constraints.

Exercise 8.4

Consider a standard LP that has an optimal basic solution. Let x_j be a non-basic variable. Assume that the reduced cost of x_j is positive. Denote by N^j its corresponding column and by N_{kj} an entry in this column. The following question addresses sensitivity analysis with respect to these entries.

1. Argue that this optimal basic solution stays feasible for any change in all entries in N^j.
2. What is the allowable increase in N_{kj} that keeps this basic solution optimal? (use the notation introduced in class).
3. What is the allowable decrease?
4. Argue that at least one of the above two values is infinite. Under what conditions are both values infinte?
5. Exemplify the 100-percent rule when dealing with simultaneous changes in entries of N^j. In particular, explain why it works.

Solution:

1. As seen in Chapter 6, any optimal solution is a basic one. Hence, $x^* = (x_B, \underline{0})$, where B is a submatrix of $A = (B \mid N)$ associated with x_B and $x_N = \underline{0}$ for the non-basic variables. The solution must also be feasible: $Ax^* = b$. Consider any change in N^j, making the matrix $\tilde{A}$ the same as A but with the new corresponding column. $\tilde{A}x^* = x_B B^{-1} + \sum_{\substack{i=1 \\ i \neq j}}^{m} x_{Ni} N^i + x_{Nj} N^j$. But since $x_N = \underline{0}$, it follows that $\tilde{A}x^* = x_B B^{-1} = b$, which makes it feasible.
2. We have shown that the optimal solution must satisfy $\bar{c}_j = c_j - c_B^T B^{-1} N_{\cdot j} \geq 0$ for all non-basic variables. This implies that prior to

any change, this condition is met. Now, consider a change Δn in the kjth entry of the matrix N. Then

$$c_j - c_B^T B^{-1}(N_{.j} + \Delta n e_k) \geq 0 \iff \bar{c}_j - \Delta n \cdot (c_B^T B^{-1})_k = \bar{c}_j - \Delta n \cdot \pi_k \geq 0.$$

For $\pi_k > 0$, the allowable increase is $\Delta n = \bar{c}_j / \pi_k$.

Similarly, for $\pi_k < 0$, the allowable decrease (when $\Delta n < 0$) is $\Delta n = c_j/(-\pi_k)$.

Note that if $\pi_k < 0$, and since $\bar{c}_j \geq 0$, the inequality remains positive for all $\Delta n \in R_+$. Therefore, the allowable increase is infinite. The same logic applies to the allowable decrease when $\pi_k > 0$. If $\pi_k = 0$, indicating that the kth constraint is not binding, both the allowable increase and decrease are infinite.

3. Consider the reduced cost in the case of simultaneous changes in the entries of N^j:

$$c_j - c_B^T B^{-1}(N^j + \Delta N^j) = \bar{c}_j - \pi \cdot \Delta N^j.$$

The requirement that these value must be non-negative implies that $\bar{c}_j \geq \sum_{i=1}^{m} \pi_i \Delta N_i^j$, which in turn implies that

$$1 \geq \sum_{i=1}^{m} \frac{\pi_i \Delta N_i^j}{\bar{c}_j} = \sum_{i=1}^{m} \frac{\Delta N_i^j}{\theta_i},$$

where θ_i represents the allowable increase (or decrease, corresponding to the sign of π_i). In other words, the current basis remains optimal if and only if the sum of all the relative changes in the entries of N^j, as a proportion of their allowable changes, is less than 1. This condition means that the 100-percent rule holds.

Bibliography

[1] Anily, S. and M. Haviv, "The price of anarchy in loss systems," *Naval Research Logistics*, vol. 69, pp. 689–701, 2022, https://doi.org/10.1002/nav.22041

[2] Bazaraa, M.S., J.J. Jarvis and H.D. Sherali, *Linear Programming and Network Flows*, 2nd edition, John Wiley and Sons, New York, 1990.

[3] Beck, A., *Introduction to Nonlinear Optimization: Theory, Algorithms, and Applications with MATLAB*, MOS-SIAM Series on Optimization, SIAM, Philadelphia, 2014.

[4] Bell, C. and S. Stidham, "Individual versus social optimization in the allocation of customers to alternative servers," *Management Science*, vol. 29, pp. 831–839, 1983.

[5] Bertsekas, D.P., *Nonlinear Programming*, Athena Scientific, 2016.

[6] Bertsekas, D.P. and J.N. Tsitisiklis, *Introduction to Probability*, 2nd edition, Anthena Scientific, 2008.

[7] Bland, R.G. "New finite pivoting rules for the simplex method," *Mathematics of Operations Research*, vol. 2, pp. 103–107, 1977.

[8] Boyd, B. and L. Vanderbergh, *Introduction to Linear Algebra: Vectors, Matrices and Least Squares*, Cambridge University Press, 2018.

[9] Boyd, B. and L. Vanderbergh, *Convex Optimization*, Cambridge University Press, 2009.

[10] Bronson, R., *Operations Research*, McGraw-Hill, 1962.

[11] Casella, G. and R.L. Berger, *Statistical Inference*, DUXBURY, 2002.

[12] Hassin, R. and M. Haviv, *To Queue or Not to Queue: Equilibrium Behaviour in Queueing Systems*, Kluwer Academic Publishers, Boston, 2003.

[13] Haviv, M., *Queues — A Course in Queueing Theory*, Springer, New York, 2013.

[14] Haviv, M., *Linear Algebra for Data Science*, World Scientific, 2023.

[15] Haviv, M., *Linear Programming, Network Flows and Markov Decision Problems*, in preparation.

[16] Haviv, M., *Introduction to Descriptive Statistics and to Probability*, in preparation.

[17] Kiefer, J., "Sequential minimum search for the maximum," *Proceedings of the American Mathematical Society*, vol. 4, pp. 502–506, 1953.

[18] Luenberger, D.G. and Y. Ye, *Linear and Non-Linear Programming*, Springer, 2008.

[19] Mood, A.M., F.A Graybill and D.C. Boes, *Introduction to the Theory of Statistics*, 3rd edition, MaGraw-Hill, 1974.

[20] Papadimitriu, C.H. and K. Steiglitz, *Combinatorial Optimization: Algorithms and Complexity*, Prentice-Hall, Inc., Englewood Cliffs, New Jersey, 1982.

[21] Sposito, V.A., *Linear Programming with Statistical Applications*, Iowa State University Press, Ames, 1989.

[22] Stidham, S., Jr. *Optimal Design of Queueing System*, CRC Press, Boca Raton, 2009.

[23] Winston, W.L., *Operations Research: Applications and Algorithms*, 4th edition, Belmont, California, 2004.

Index

Affine function, 5, 62, 107, 124, 162, 198, 272
Aitken acceleration, 28
Arithmetic mean, 6, 12, 56, 166, 179, 239, 276, 277
Artificial variable, 210, 212, 214

Basic solution, 189, 191, 192, 194, 197, 198, 200, 202, 203, 206, 207, 210, 218, 236, 298
Basis, 189
Bernoulli distribution, 16
Binary search, 33

Caratheodory theorem, 220, 286
Cauchy–Schwarz inequality, 14, 63, 258
Chernoff bound, 14–16, 241
Cholesky factorization, 77
Class-dominance, 105
Column generation, 204, 214
complementary slackness conditions, 159, 276
Concave function, 8, 12, 18, 19
Condition number, 69
Contour level, 131
Convex function, 8, 10–12, 30, 31, 40, 44–46, 96, 99
Convex set, 40, 45, 162, 192, 219
    Convex combination, 42, 47, 204, 219–221, 281, 286, 287, 290

Extreme point, 192, 193, 219–221, 244, 287, 288, 290
Convex-hull, 42, 220

Degeneracy, 199, 218, 231
Duality, 144
    Dual function, 144
    Dual problem, 144
    Duality in linear programming, 149
    Duality gap, 145, 148
    Weak duality, 144, 146

Eigensystems
    Algebric multiplicity, 124
    Geometric multiplicity, 124
Envelope theorem, 141

Feasible path, 131
First–order conditions (FOCs), 5, 50, 51, 115, 118
Fitted value, 56
Fixed point, 85, 87

Gauss–Markov theorem, 129
Geometric mean, 7, 12, 239
Golden search, 34
Gradient descent, 64
Gradient inequality, 43

Hölder inequality, 14
Harmonic mean, 7, 13, 97, 239

www.ingramcontent.com/pod-product-compliance
Lightning Source LLC
Chambersburg PA
CBHW052119230326
41598CB00080B/3884